Chemical Aspects of Drug Delivery Systems

Chemical Aspects of Drug Delivery Systems

Edited by
D. R. Karsa
Akcros Chemicals UK Ltd., Manchester

R. A. Stephenson
Chemical Consultant

THE ROYAL
SOCIETY OF
CHEMISTRY
Information
Services

The proceedings of a Symposium organized by the Waterborne Polymers Group of
The British Association for Chemical Specialities (BACS) and the MACRO Group UK
(Joint Group of The Royal Society of Chemistry and the Society of Chemical Industry)
at Salford University on 17–18 April 1996.

Special Publication No. 178

ISBN 0-85404-706-9

A catalogue record for this book is available from the British Library.

Published by The Royal Society of Chemistry,
Thomas Graham House, Science Park, Milton Road,
Cambridge CB4 4WF, UK

Printed by Bookcraft (Bath) Ltd

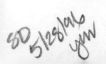

Introduction

Despite the advances in the development of new drugs, a drug may never reach the target organ, or it may be difficult to achieve the necessary level of drug in the body. Large doses can result in serious side effects, and can harm normal cells and organs as well as diseased cells. Hence controlled release and the targeting of delivery systems must evolve in parallel to drug research.

This symposium, jointly organized by the Waterborne Polymers Group of BACS (The British Association for Chemical Specialities) and Macro Group UK (the joint Group of the Royal Society of Chemistry and the Society of Chemical Industry), covers some of the advances in the Chemical Aspects of Drug Delivery Systems.

New materials for drug delivery and targeting are reviewed and a representative range of excipients and delivery systems is considered in depth. Particular attention is given to poly(ethylene oxides) and derivatives, carbohydrate derivatives (including starch, lactose and microcrystalline cellulose) and selected water-soluble polymers and hydrogels.

Although a single volume can never cover all aspects of so broad a topic, the editors hope that this volume will serve as a useful introduction to chemists and pharmacists new to this field of research and a valuable addition to those who are already familiar with this subject.

R. A. Stephenson
D. Karsa

Contents

New Materials and Systems for Drug Delivery and Targeting

P. York

POSTGRADUATE STUDIES IN PHARMACEUTICAL TECHNOLOGY, SCHOOL OF PHARMACY,
UNIVERSITY OF BRADFORD, BRADFORD BD7 1DP, UK

1 INTRODUCTION

The range of bioactive substances emerging as potential drug candidates, together with those currently under research and development, continue to provide major challenges for efficient drug delivery and targeting. Various strategies for formulation design using diverse chemicals as formulation excipients are available, and numerous materials are being considered and developed to provide specific functionalities in the design of medicines. A number of approaches to formulation and drug delivery will be discussed at this symposium and this lecture, serving as a general introduction, highlights the need and value of these approaches coupled with the related issues of excipient properties and design.

It has been a constant ambition of formulation scientists to optimise drug delivery systems which provide a defined dose, at a chosen rate, at a selected time, to a targeted biological site. Whilst improvements in drug delivery over recent years are impressive, there is still some way to go in fully achieving these objectives. Key issues requiring continuing research and study range from fundamental understanding of the biosystems and targets and basic characterisation of novel classes of bioactive agents, to the development of 'designer' or 'smart' materials which provide required excipient or carrier properties to achieve modulated and targeted drug delivery. Coupled with these activities is the necessary realism of the practical constraints imposed in designing drug delivery systems. These include the necessity of using materials which will achieve regulatory approval and clearance, and the constraints imposed by the nature of the various routes of administration available for drug delivery.

2 ROUTES OF ADMINISTRATION AND CLASSIFICATION OF DRUG DELIVERY SYSTEMS

The principal routes of administration for medicinal products are listed in Table 1, together with a general classification of the main groups of traditional dosage forms. The choice of an appropriate route of administration for a specific bioactive will be influenced by many factors, such as required time of onset of action or drug targeting issues. Similarly, selection of drug delivery class is based on these and other numerous factors, as well as features related to the properties of the bioactive material itself, such as solubility and stability.

Table 1 Routes of administration and general classification of drug delivery systems

ROUTES OF ADMINISTRATION	DRUG DELIVERY CLASS
ORAL	SOLIDS (e.g. tablets, capsules)
PULMONARY	SEMI-SOLIDS (e.g. gels, creams)
TOPICAL	LIQUIDS -
	SOLUTIONS
BODY CAVITY (e.g. nasal, eye, buccal)	COLLOIDS
PARENTERAL (e.g. i/v, i/m, s/c)	EMULSIONS

The explosion of synthetic and semi-synthetic bioactive substances in the 1950's and 1960's, which continues to the present day, led to the development of a range of the conventional dosage forms which dominate the range of medicines available today. However, newer trends and strategies in drug discovery with the advent of highly potent compounds or those requiring location at specific biological tissues or sites has led to the development of alternative drug delivery systems, which attempt to address the requirements of rate and extent of drug release, and thereby absorption. Delivery systems include oral sustained release formulations[1] (e.g. multiple unit disintegrating particles or beads, single unit non-disintegrating system), controlled release preparations (e.g. oral osmotic pump[2]) and bioadhesives[3] and liposomes[4]. The products of biotechnology research in the 1980's and 1990's have imposed even greater demands on drug delivery formulations and drug targeting with the emergence of peptides, problems, oligonucleotides and elements of DNA as potential drug candidates, since specific challenging features of such bioactives in terms of efficient and safe drug delivery need to be addressed from the points of view of administration route and suitable excipient and carrier materials.

Table 2 highlights the various groups of chemicals that are used as vehicles, carriers and excipients in both conventional and more recent approaches to formulating medicines. Much research activity is focused on the development and testing of new carrier systems, such as biodegradable polymers, such as the polylactides, and composite materials like low density lipoproteins.

3 DRUG DELIVERY AND TARGETING

A diagramatic illustration of the inter-relationship between the components controlling the processes of drug delivery and targeting is presented in Figure 1. In the diagram, the

Table 2 Range of chemical groups used as vehicles, carriers and functional excipients in drug delivery systems

CHEMICAL GROUP	EXAMPLES	
INORGANICS	- CALCIUM PHOSPHATE	- diluent for solid dosage forms
	- TITANIUM DIOXIDE	- opacifying agent
CARBOHYDRATES	- α - LACTOSE MONOHYDRATE	- direct compression tableting excipient
	- β - CYCLODEXTRIN	- drug complexing agent
SURFACTANTS	- SODIUM LAURYL SULPHATE	- promote drug particle wetting
	- POLOXAMERS	- targeting of colloidal particles
POLYMERS	- STARCH	- tablet disintegrant
	- ETHYL CELLULOSE	- film former for coating solid dosage forms
	- HYDROGELS	- matrix for controlled drug release
	- POLYACRYLIC ACIDS	- bioadhesive polymers
LIPIDS, FATS	- POLYLACTIDES	- biodegradable polymer
	- STEARIC ACID	- tablet lubricant
	- GLYCERIDES	- matrix for controlled drug release
	- PHOSPHOLIPIDS	- formation of liposomes
AMINO-ACIDS, PEPTIDES, PROTEINS	- LEUCINE	- tablet lubricant
	- LOW DENSITY LIPOPROTEIN	- microparticulate drug carrier system

drug is delivered via a carrier system and four situations are identified. In A, leakage of the drug occurs as the drug-carrier system moves down the route of administration, whilst in B the drug is lost via the walls of the delivery route to non-target sites. In C and D, successful drug delivery to the target sites is achieved although drug loss to surrounding tissues can occur. Careful attention must thus be given to the characteristics of the targets, including access and location, as well as the characteristics of any carrier materials incorporated into drug delivery systems.

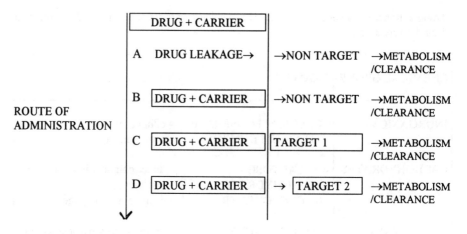

Figure 1   *Components for drug delivery and targeting - route of administration, carrier and target (adapted from Florence[4])*

4 CARRIER SYSTEMS

In many cases, carrier materials are used in particulate forms, and Table 3 lists various types of microparticle colloidal carrier systems, together with ranges of particle diameters in nanometers.   Microspheres and nanoparticles have continuous matrices containing dispersed or dissolved drug whilst microcapsules and nanocapsules are composed of a drug core surrounded by a layer acting as a coating or barrier to drug diffusion or dissolution.   Vesicles are made up of single or multi-lamellar bilayer spherical particles containing drug within their lipid or aqueous regions.   Emulsion and microemulsions are composed of oil or aqueous droplets dispersed in a continuous phase of the other liquid, or multiple emulsions (ie, oil-in-water-in-oil and water-in-oil-in-water[5]), with drug dissolved in either or both oil and aqueous phases.   Low-density lipoproteins have the benefit of being natural materials and drug can, for example, be adsorbed onto the protein or phospholipid head groups, solubilised in the lipid containing core, or attached to the surface.   The range in particle sizes available for the various carrier systems provides potential regarding choice of administration route allowing smaller particles to be administered by parenteral routes for intravenous, subcutaneous and intramuscular drug delivery.

The types of carrier materials used, the drug substance and the biological environment for drug delivery all influence the mechanisms of drug release.   Table 4 highlights the principal release mechanisms and drug, particle and environmental factors influencing drug release.   The complex matrix of variables and interactions which influence and ultimately control drug release will clearly continue to provide major challenges for pharmaceutical scientists working in drug delivery and targeting.

Table 3 Microparticulate colloidal carrier systems*

SYSTEM	SIZE (nanometers)
MICROCAPSULES	1000-100,000
EMULSIONS	200 - 30,000
MICROSPHERES	200 - 30,000
MULTI-LAMELLAR VESICLES	200 - 3,500
NANOCAPSULES	50 - 200
UNI-LAMELLAR VESICLES	25 - 200
NANOPARTICLES	25 - 200
MICROEMULSIONS	20 - 50
LOW-DENSITY LIPOPROTEINS	20 - 25

(*adapted from Florence[4])

5 FORMULATION STRATEGIES FOR CONTROLLED DRUG RELEASE AND DRUG TARGETING

A variety of approaches to formulation design are available and are being developed, some of which incorporate the newer excipients and materials with specific and directed functionality in terms of drug release. Tables 5A and 5B list a number of formulation systems used by the oral, parenteral and pulmonary routes. Strategies range from chemical modification of the drug substance to provide a lower solubility salt, to more complex drug delivery systems involving enzymatic breakdown of a formulation component or particle coating to effect drug release[7], to the delivery of drugs containing liposomes to the lungs by nebulisation[8].

Recent developments have further extended opportunities with the advent of externally activated drug delivery systems (see Table 6). Activating sources based on heat, sound, light, electrical pulses and magnetic fields are coupled with advanced materials incorporated into dosage forms to achieve controlled, pulsed and/or modulated drug release. Whilst many of these systems are in their infancy the potential of these approaches will continue to be explored, undoubtedly leading to advanced drug delivery systems.

Table 4 Mechanisms and properties influencing drug release from particle carrier systems*

MECHANISM OF DRUG RELEASE	PROPERTIES INFLUENCING DRUG RELEASE	
DIFFUSION	DRUG	- concentration, particulate location and distribution
		- molecular weight; physicochemical properties
EROSION		- drug:carrier interactions
	PARTICLE	- type, amount adjuvants, matrix
PARTICLE DISINTEGRATION		- size, density, surface properties
CHARGE EFFECTS	BIOLOGICAL FLUID/ ENVIRONMENT	- pH, polarity, ionic strength
		- surface tension, viscosity
		- temperature
MEMBRANE PERMEABILITY CONTROL		- enzyme

(*adapted from Tomlinson[6])

6 DRUG PARTICLE ENGINEERING

Drug particle engineering, or crystal engineering, provides an additional dimension to drug delivery and targeting. Traditional methods of particle formation, crystallisation and precipitation from solvents, do not generally provide the preferred properties required for formulation and processing of drugs into drug delivery systems, and it is common for additional processing to be carried out, such as milling and classification. However whilst such extra processing provides desired characteristics (eg, particle size and size distributions), changes in other properties can take place in an uncontrolled manner leading to batch inconsistency and thereby lack of precise control of performance in formulated products.[9] Physicochemical changes observed include solid state phase transitions and surface crystallisation. In this respect the non-equivalency of particles resulting from conventional crystallisation, harvesting and drying operation can be added to by further processing. Understanding of these changes has been facilitated by recent developments in high resolution analytical techniques, such as microcalorimetry[10] inverse phase gas chromatography[11] and x-ray powder diffraction[12]. The concept of optimising particulate formulations in terms of surface properties, such

Table 5A   Formulation systems for controlled drug release and drug targeting

ORAL ROUTE

ADMINISTRATION ROUTE	FORMULATION SYSTEM	
ORAL - CONTROLLED RELEASE	MODIFICATION OF DRUG SUBSTANCE	eg low solubility salt eg drug polymer complex

	MODIFICATION OF DELIVERY SYSTEM
	  - MULTIPLE UNIT
	  - SINGLE UNIT
		
		eg coated pellets
		eg osmotic pump

ORAL - DRUG TARGETING	BIOADHESIVE SYSTEMS	eg buccal mucoadhesive formulations
	PHYSIOLOGICAL/ PHARMACOLOGICAL CONTROL	eg enteric coating eg release of locally active excipients
	PRODUCT DENSITY COLON SPECIFIC DELIVERY	eg gastric floating systems eg microbiologically controlled delivery

	PULSATILE DRUG DELIVERY
	  - PHYSICAL
	
	  - BIOLOGICAL
		
		eg swelling of hydrogel
		      'plug' to release pulse
		      of drug
		eg drug release after
		      enzymatic breakdown
		      of coating

as surface energy requirements for powder inhalation drug delivery systems, is becoming a practical reality.

The attractive alternative approach of producing drug particles and crystals with desired properties, such as particle size, shape, surface-free energy and crystallinity has been realised through the use of super-critical fluid technology[13]. In the SEDS process (Solution Enhanced Dispersion by Supercritical Fluids), two streams with one composed of a liquid solution containing the drug and a second containing supercritical carbon dioxide are introduced simultaneously using a coaxial nozzle arrangement into a particle formation vessel held at constant temperature and pressure supercritical conditions. The process involves virtually instantaneous dispersion, mixing and extraction of the solution solvent by the supercritical fluid leading to very high supersaturation ratios. These

Table 5B   Formulation systems for controlled drug release and drug targeting

PARENTERAL AND PULMONARY ROUTES

ADMINISTRATION ROUTE	FORMULATION SYSTEMS	
INTRAVENOUS	MIXED SOLVENTS	
	EMULSIONS/ MICROEMULSIONS	eg lipid emulsions
	COLLOIDAL SYSTEMS	eg liposomes
		eg mixed micellar systems
INTRAMUSCULAR/ SUBCUTANEOUS	BIODEGRADABLE FIBRES	eg contraceptive implant
	FIBRIN-GELS	eg fibrin -antibiotic mixtures
	BIODEGRADABLE POLYMERS	eg polylactides
PULMONARY	INHALERS	eg metered dose inhalers
		eg powder inhalers
	NEBULISERS	eg liposomes

factors, together with precise control of relative flow rates of the two streams in the nozzle, provide uniform conditions for nucleation and particle formation, and the pure, solvent-free product is retained in the particle formation vessel. By varying the working conditions and changing the drug solution solvent, it has been shown possible to provide directed control over particle size, shape morphology, purity and polymorphic form. This capacity provides benefits over other reported techniques for particle formation using supercritical fluids[14,15] and clearly such precise manipulation of critically important properties of drug and carrier particles, coupled to consistency within and between batch, provides vast opportunities for drug delivery and targeted systems.

7 CONCLUDING REMARKS

Whilst conventional dosage forms, such as tablets and hard gelatin capsules, composed of drugs with traditional excipients, continue today as the vast majority of formulations available for drug administration, major progress has been achieved over recent years in the fields of controlled drug delivery and targeting. Success in these areas is important both to improve the bioperformance and efficiency of drug delivery systems and to deal with recent trends in drug discovery. The range of materials used as functional excipients and carriers continues to grow, as does the novelty of alternative approaches in drug targeting. Nevertheless, the therapeutic agents emerging from studies in biotechnology, such as proteins and gene constructs, demand further research and

Table 6 Externally activated drug delivery systems

EXTERNAL ACTIVATION	FORMULATION SYSTEM	MECHANISM FOR DRUG DELIVERY CONTROL
HEAT	LIPOSOMES	CHANGE IN PERMEABILITY
	HYDROGELS	CHANGE IN SWELLING
ULTRA-SOUND	POLYMERS	CHANGE IN PERMEABILITY
ELECTRICAL - PULSE	GELS	CHANGE IN PERMEABILITY, SWELLING
- IONTOPHORESIS	TRANSDERMAL SYSTEM	CONTROL LOCATION AND DURATION OF DRUG RELEASE
- ELECTROPHORESIS	ERODIBLE GELS	CHANGE IN PHYSICAL FORM (SOLID TO SOLUTION) IN ELECTRIC FIELD
MAGNETIC MODULATION	MAGNETICALLY RESPONSIVE MICROSPHERES CONTAINING (eg) Fe_3O_4	MAGNETIC FIELD CAN RETARD FLUID FLOW OF PARTICLES
LIGHT	PHOTORESPONSIVE HYDROGELS CONTAINING AZO-DERIVATIVES	CHANGE IN DIFFUSIONAL CHANNELS, ACTIVATED BY SPECIFIC WAVELENGTH

creativity due to their particular properties and targeting requirements. All these developments need to be paralleled by research and inventiveness in pharmaceutical material science and control of particles during their formation. Recent advances in applying supercritical fluid technologies do, however, provide opportunities for future developments in these areas.

References

1. J. R. Robinson and V. H. L. Lee, (eds) 'Controlled Drug Delivery: Fundamentals and Applications', Marcel Dekker, New York, 1987.
2. G. Santus, R. W. Baker, *J.Contr.Rel.*, 1995, **36**, 1.
3. M. Veillard, 'Bioadhesion Possibilities and Future Trends', Wiss, Stuttgart, 1991.
4. A. T. Florence, 'Drug Delivery: Advances and Commercial Opportunities', Connect Pharma, Oxford, 1994.
5. A. T. Florence, T. L. Whateley and J. Omotosho, 'Controlled Release of Drugs: Polymers and Aggregate Systems', VCH, New York, 1989.
6. E. Tomlinson, 'Drug Delivery Systems', Ellis Horwood, Chichester, 1987.
7. A. Rubenstein, *Biopharm. Drug Dispos.*, 1990, **11**, 465.
8. K. M. G. Taylor and S. J. Farr, 'Liposomes in Drug Delivery', Harwood, Chichester, 1993.
9. P. York, *Int.J.Pharm.*, 1993, **14**, 1.
10. G. Buckton, 'Excipients and Delivery Systems for Pharmaceutical Formulation', Royal Society of Chemistry, London, 1995.
11. M. Ticehurst, R. C. Rowe and P. York, *Int.J.Pharm.*, 1994, **111**, 241.
12. M. Landin, R. C. Rowe, P. York, *Eur.J.Pharm.Sci*, 1994, **2**, 245.
13. M. H. Hanna, P. York, D. Rudd, S. Beach, *Pharm.Res.*, 1995, **12**, S141.
14. J. W. Tom, P. G. Debendetti, *J.Aerosol Sci.*, 1991, **22**, 555.
15. R. Bodmeier, H. Wang, D. J. Dixon, S. Mawson and K. P. Johnston, *Pharm.Res.*, 1995, **12**, 1211.

The Use of Bioadhesive Polymers as a Means of Improving Drug Delivery

Slobodanka Tamburic and Duncan Q. M. Craig

CENTRE FOR MATERIALS SCIENCE, SCHOOL OF PHARMACY, UNIVERSITY OF LONDON, 29–39 BRUNSWICK SQUARE, LONDON WC1N 1AX, UK

1 INTRODUCTION

Bioadhesion is a complex phenomenon related to the ability of some natural and synthetic macromolecules to adhere to biological tissues. In medical applications, bioadhesion has been employed in surgery and dentistry for many years through the use of "super glues", particularly the esters of α-cyanoacrylates, polyurethanes, epoxy resins, acrylates and polystyrene[1]. The mechanism of bonding in these cases usually involves the formation of covalent bonds with the target tissue (bond or tooth), providing a permanent linkage.

If the biological substrate is a mucus membrane, bioadhesive interactions occur primarily with the mucus layer and this process is referred to as mucoadhesion. The bonds involved are more likely to be of secondary chemical nature, combined with physical entanglement of polymer chains. The process is a reversible one, where the detachment of the mucoadhesive is caused either by the breakage of low energy bonds or by the physiological process of mucus turnover.

Pharmaceutical applications of bio(muco)adhesion have been the subject of great interest and intensive research during the last decade. Bioadhesive polymers fulfil the following desirable features of a controlled release system[2]: a) localisation in specified regions to improve and enhance bioavailability of drugs b) optimum contact with the absorbing surface to permit modification of tissue permeability, which is especially important in the case of peptides/proteins and ionised species, and c) prolonged residence time to permit once-daily dosing, thus improving patient compliance. To date, the use of mucoadhesion in prolonging and controlling drug delivery has been employed with respect to a number of mucus membranes, i.e. gastrointestinal, ocular, nasal, oral, vaginal and rectal. Theoretically, mucoadhesion could resolve several problems of controlled release drug delivery systems, particularly the low availability of some drugs, short residence time, first-pass metabolism and insufficient patient compliance. Both topical and systemic administration of active agents has been studied using a wide variety of dosage forms, including tablets, patches, films, discs, ointments, gels, powders, beads, microcapsules, liposomes and plasters.

This paper will describe some aspects of bioadhesion, such as mucus structure, stages of adhesion and the theories proposed to explain the phenomenon. A range of bioadhesive polymers have been examined so far, and these will be reviewed, along with the factors that affect the bioadhesive strength, the testing techniques used and the dosage forms studied. In addition, some results of our work, focused on the use of poly(acrylic acid) polymers, will be presented.

2 THE MUCUS LAYER

Mucus is a continuous layer covering all the internal tracts of the body and having both a protective and lubricating role. It is a gel-like structure that adheres firmly to the epithelial cell surface. In most cases, the adhesive interaction would initially be between the bioadhesive polymer and the mucus layer, and would not directly involve the epithelial surface[3]. Since an understanding of the target tissue is essential in considering the interactions with mucoadhesive polymers, a brief review on mucus structure and properties will be given.

Mucus is a mixture of large glycoproteins (mucins), water, electrolytes, sloughed epithelial cells, enzymes, bacteria and bacterial products and various other materials, depending on its source and location. The main components of mucus are mucin glycoproteins (less than 5% of the total weight), which are responsible for its rheological, adhesive and cohesive properties. Mucin glycoprotein chemically consists of a large peptide backbone with pendant oligosaccharide side chains, many of which terminate in either sialic or sulphonic acid[4] or L-fucose[5] groups. The oligosaccharide chains are covalently linked to the protein core by O-glucidic bonds, essentially between N-acetylgalactosamine and serine or threonine[6]. About 25% of the polypeptide backbone is without sugars but rich in charged amino acids, especially aspartic acid. This region is involved in cross-linking via disulphide bonds between mucin molecules[7]. A highly extended and flexible conformation is suggested for mucin glycoproteins to permit maximum water sorption (more than 95% of the total weight). The mucin molecule behaves as an anionic polyelectrolyte at physiological pH, since terminal sialic acid groups have a pK_a value of 2.6[8], with sulphate residues contributing equally to the negative charge.

The mucus gel structure is a consequence of the intermolecular association of glycoproteins in the polymeric network. The mucin molecule is believed to be a terminally linked chain with numerous cross-linkages[9] (Figure 1). The entangled nature of mucus is due to disulphide linkages, physical entanglements and secondary bonds, i.e. electrostatic and hydrogen bonding and hydrophobic interactions[10].

A   entanglements
B   molecular associations
C   permanent crosslink

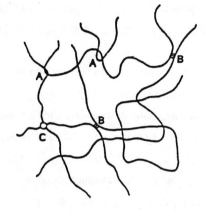

Figure 1. Cross-linked structure of the mucus network

A proportion of the glycoprotein is not incorporated in the network but is present as a soluble fraction, enhancing the viscosity of the interstitial fluid[11]. The adjacent epithelial layer with "fuzzy coat" glycocalyx largely contributes to the formation of a strong tissue adhesion with the mucus layer.

Based on the structure of mucin, there are four properties of the mucus layer that may relate to mucoadhesion[1]:

1) It is a network of linear, flexible and random coil mucin molecules

2) It is negatively charged

3) It is a cross-linked network and

4) It is highly hydrated.

3 MECHANISMS AND PRINCIPLES OF MUCOADHESION

Ghandi and Robinson[3] considered the formation of an adhesive bond between a polymer and biological tissue as a two-step process. The first step involves the initial contact between the two surfaces, while the second involves the formation of secondary bonds due to non-covalent interactions. Duchêne et al[11] have suggested that for bioadhesion to occur a succession of phenomena is required, whose role depends on the bioadhesive polymer nature. They summarise the stages of bio(muco)adhesion as follows:

a) An intimate contact between the bioadhesive polymer and the receptor tissue must exist, resulting either from a good wetting of the bioadhesion surface or the swelling of the polymer.

b) The penetration of the polymer into the tissue surface or the interpenetration of the polymer and mucin chains.

c) The formation of secondary chemical bonds.

There are a large number of adhesion theories that have been applied to mucoadhesion in an attempt to describe and understand this complex process, including wetting, diffusion, electrostatic, fracture and adsorption theories[3,11]. In addition, some authors have proposed theories that are a combination of several approaches, such as the adsorption/interdiffusion[12] (Figure 2) or fracture/interpenetration theory[13].

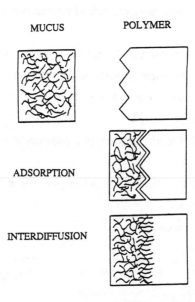

Figure 2. Schematic representation of adsorption/interdiffusion theory

A factor that may maximise contact between the mucoadhesive and mucus (which is undoubtably desirable in achieving good mucoadhesion) is interpenetration or interdiffusion of mucus and the polymer. This is in turn influenced by such factors as polymer chain mobility, chain entanglement, cross-linking density, equilibrium swelling, porosity, presence of additives and compatibility of the two surfaces[1].

Covalent bonds are uncommon with mucoadhesion, while secondary chemical bonds involved in this process comprise electrostatic and hydrophobic interactions, hydrogen bonding and van der Waals intermolecular interactions. For charged bioadhesives, electrostatic and hydrogen bonds are of primary importance. Interestingly, negatively charged polymers (eg. polyacrylic resins) are established as good mucoadhesives, although possessing the same charge as mucus. It is known, however, that two surfaces may attract each other through long-range forces created by atomic and molecular vibrations that produce fluctuating dipoles on each surface[1]. Similarly, hydrogen bonding is experimentally proven to be very important in the process of mucoadhesion [14,15], while hydrophobic bonding, which takes place between non-polar groups in aqueous solution, is also considered to play a significant role in bio(muco)adhesion[9].

The most effective mucoadhesives are found to be linear or lightly cross-linked polymers which differ considerably in structure to the mucus glycoprotein molecules, hence it is unlikely that they adhere through interactions similar to mucin-mucin interactions. It is conceivable that penetration takes place between oligosaccharide side chains on the mucus and the "free ends" of the interacting polymers[2]. Since good wetting and spreading are necessary to guarantee molecular contact between the two phases, the surface characteristics of both bioadhesives[16,17] and mucin solutions[18] in terms of contact angle, spreading coefficient and surface free energy have been studied. Mortazavi and Smart[19] have proposed mucus gel dehydration and intermolecular complex formation as important factors in gel strengthening during mucoadhesion.

4 MUCOADHESIVE POLYMERS

The development of mucoadhesive polymers can be traced back as far as 1947, when gum tragacanth and dental adhesive powders were combined to form a vehicle for applying penicillin to the oral mucosa[20]. An improvement of this system was achieved by the combination of carboxymethyl cellulose (CMC) and petrolatum, followed by the formulation of "Orahesive" (a powder mixture of Sodium CMC, pectin and gelatin) and "Orabase" (the same mixture dispersed in a polyethylene/mineral oil base)[21]. The next step was the blending of SCMC with poly(isobutylene) (PIB) and laminating this mixture onto a polyethylene sheet. One of the first comprehensive analysis of bioadhesive strength of different polymers was performed by Chen and Cyr[22]. The authors had tested a range of the SCMC/PIB blends, characterising their adhesiveness in terms of "excellent, satisfactory, fair and poor". The polymers identified as "excellent" were; sodium alginate, SCMC, Guar gum, hydroxyethyl cellulose (HEC), Karaya gum, methyl cellulose (MC), poly(ethylene glycol) (PEG) and gum tragacanth[22]. Acrylic polymers (the homo- and co-polymers of acrylic acid and its esters) were soon identified as very good mucoadhesives and extensively investigated. A large number of patents (see [2]) deal with the blends of poly(acrylic acid) (PAA) and either hydroxypropyl cellulose (HPC) or MC in mucoadhesive preparations. The most studied mucoadhesives through the 1980s have been PAA, HPC and SCMC.

An ideal mucoadhesive polymer has to be non-toxic, non-absorbable from the GI tract, capable of forming strong non-covalent bonds with mucin/epithelial cell surfaces, it should adhere quickly to moist tissue, allow easy incorporation of drug and its controlled release, possess specific sites of attachment and be economical[23]. Pharmaceutical scientists have been using various approaches in the search for an ideal mucoadhesive, including the chemical modification of existing adhesives, the synthesis of novel polymers and the combination of a number of mucoadhesive polymers.

Investigations into polymers with various molecular characteristics, conducted by many authors (eg.[24,25]), have led to a number of conclusions regarding the molecular characteristics required for mucoadhesion. The properties exhibited by a good mucoadhesive may be summarised as follows[12]:

    a) Strong hydrogen bonding groups (-OH, -COOH)

    b) Strong anionic charges

    c) Sufficient flexibility to penetrate the mucus network or tissue crevices

    d) Surface tension characteristics suitable for wetting mucus/mucosal tissue surfaces

    e) High molecular weight.

Although an anionic nature is preferable for a good mucoadhesive, a range of nonionic molecules (eg. cellulose derivatives) and some cationic (eg. chitosan) can be successfully used. A short list of mucoadhesive polymers is given below:

A. SYNTHETIC POLYMERS

- Cellulose derivatives (MC, EC, HEC, HPC, HPMC, SCMC)
- poly(acrylic acid) polymers (carbomers, polycarbophil)
- poly (hydroxyethyl methylacrylate)
- poly(ethylene oxide)
- poly(vinyl pyrrolidone)
- poly(vinyl alcohol)

B. NATURAL POLYMERS

- Tragacanth
- Na alginate
- Karya gum
- Guar gum
- Xanthan gum
- Lectin
- Soluble starch
- Gelatin
- Pectin
- Chitosan

The high proportion of bioadhesive polymers that contain carboxyl groups is in accordance with the suggestion that secondary bond formation is the principal source of mucoadhesion, as unionised carboxyl groups are able to form strong hydrogen bonds, while in the ionised form they may interact electrostatically. The effect of other secondary bond-forming groups (e.g. hydroxyl, ether oxygen, amine, amide) on the mucoadhesive properties is not as well defined as for the -COOH group[2].

The factors which affect bioadhesion are defined by both the nature of the polymer and environmental conditions. The polymer-related factors are: molecular weight and molecular conformation, cross-linking density, charge and ionisation, concentration of the polymer and its swelling characteristics. The surrounding medium contributes via pH, ionic strength and the nature of the dissolved ions, all of which may affect the polymer hydration, among other factors, especially in the case of charged bioadhesives. The strength and duration of application of the polymer on the substrate are additional important factors in establishing mucoadhesive bonds. In general, it may be concluded that polymers with flexible chains that form an expanded network and are compatible with mucin are favourable candidates for use as adhesives in developing bioadhesive dosage forms.

Quantification of adhesive bond strength is necessary in screening and developing mucoadhesive polymers, as well as in the formulation of drug delivery systems. No standard method has been proposed for this purpose so far, resulting in a variety of techniques being used by different authors. Consequently, the data obtained are often difficult to compare and are sometimes controversial. The most usual classification of the methods for measurement of mucoadhesion is *in vitro* methods (which often require an artificial biological medium such as mucus or saliva) and *in vivo* methods. Some authors distinguish a category of *ex vivo (in situ)* methods, where an animal tissue is used under the controlled conditions. Most of the techniques used have been designed to measure tensile strength[25,26,27], but some also deal with peel[28,29] and shear strength[30]. A large number of mucus-simulating media have been employed, including crude or purified mucin, "homogenised" mucin, wet dialysing membrane, hydrated polyvinyl pyrrolidine/cellulose acetate hydrogel etc.. The animal tissues generally studied include porcine oesophagus and peritoneum, isolated buccal mucosa, rabbit gastric mucosa and mouse peritoneal membrane. The most widely used parameters in quantifying mucoadhesiveness are maximum detachment force and work of adhesion, although the duration of adhesion has recently gained considerable attention[31].

Some authors have approached the problem differently, such as Robinson's group[24] who developed a fluorescence probe technique using cell cultures which indirectly measures the binding between a polymer and epithelial cells. Another interesting approach is the use of a rheological method, established by Hassan and Gallo[32], whereby the rheological properties of mucin/polymer mixtures are related to the mucoadhesiveness of the polymer. A recent study by the group of the University of Pavia[33] has shown a good correlation between rheological and tensile parameters in the case of SCMC. Detailed reviews on the experimental methods for determination of bioadhesive strength can be found elsewhere[2,11,12,23,34].

5 BIOADHESIVE DOSAGE FORMS

Bio(muco)adhesive dosage forms are a relatively new type of preparation that may be used to treat both local and systemic diseases. In terms of the site of application, there are three major groups of bioadhesive delivery systems, namely oral, parenteral and local (topical) administration. The main benefits expected from orally applied mucoadhesive dosage forms, especially those containing peptides, are[35]

a) prolonged residence time at the site of drug absorption (eg. by controlling GI transit)

b) increased contact with the absorbing mucosa, resulting in a steep concentration gradient to favour drug absorption

c) localisation in specified regions to improve and enhance the bioavailability of the drug (e.g. targeting to the colon).

The carrier systems used are often colloidal in nature, including liposomes, nanoparticles, nanocapsules, microcapsules, niosomes and emulsions, coated or mixed with mucoadhesive polymers. In addition, the use of albumin beads[36], pellets[37] and even magnetic granules[38] have been examined with encouraging results. However, the complexity of the physiology of the GI tract and the large number of factors affecting its function renders it difficult to prepare reliable oral mucoadhesive dosage forms.

Parenteral bioadhesive systems are still in their infancy, although theoretically having great advantages, including targeted delivery[39]. After an i.v. application, the fate of a drug is largely determined by the action of the immune system; therefore an effective defence from the action of macrophages has to be achieved, as with all colloidal dosage forms.

Locally or topically applied bioadhesive systems can be used to achieve both local and systemic action. In terms of the site of application, there are several categories of bioadhesive dosage forms, intended for the skin (intact, diseased or wounded), oral cavity, vagina, rectum, nasal cavity and the eye. A comprehensive review of local bioadhesive delivery systems has been given by Hollingsbee and Timmins[40]. Since the delivery of drugs to the oral cavity is of particular interest for our research, a brief review of achievements in this field will be given.

5.1 Oral mucosal drug delivery systems

The accessibility of the oral cavity makes it a potentially attractive route for drug delivery. However, rapid removal of conventional delivery systems, basically through salivary flow and mouth movements, and the relative impermeability of the buccal tissue are major impediments. Ideally, bioadhesive polymers may overcome the removal issue, and could be combined with penetration enhancers to generate a novel and successful drug delivery system.

De Vries et al[40] have given the following requirements for oral adhesive dosage forms; a) they should be flexible enough to follow the movement of the mouth b) they should be adhesive enough to be retained on the oral mucosa, but not so strong that the mucosa is damaged on removal, and c) they should be biocompatible and non-irritating.

Both systemic and local drug delivery is possible through the oral mucosa. There are three main regions in the mouth, differing principally in the permeability of the mucosa and salivary flow, these being the buccal, sublingual and dental/gingival regions. Both buccal and sublingual regions are extensively used for systemic delivery, with the latter being more permeable but not very suitable for prolonged mucoadhesion due to the physical structure and mobility of the tissue. Consequently, the sublingual area is used mainly for the delivery of drugs which require a rapid onset of action such as glyceryl trinitrate (given in the form of quick-dissolving tablets or aerosols)[41]. Recently, a considerable effort has been made to develop buccal delivery of peptide and protein drugs for systemic delivery, but most of the studies are still in experimental stages[42]. Both buccal and dental/gingival region have been used for local therapy, e.g. in the case of paradontoses, aphtae and lesions by trauma, in and after dental procedures, or for the application of fluoride.

Oral bio(muco)adhesive delivery systems have been formulated mainly in the form of sustained-release tablets, semisolid preparations, films, patches, powders and plasters. The first oral mucosal delivery system, Orabase[R] (a blend of pectin, gelatin and SCMC dispersed in polyethylene/mineral oil base), is successfully used for buccal delivery of steroids for mucosal ulceration[43]. Gingival plasters containing prostaglandins ($PGF_{2\alpha}$ and PGE_2) were formulated by Nagai et al[44] to provide a continuous slow release into the gingival tissue for orthodontic tooth movement. The major components of the formulation were a synthetic resin, natural gum, hydrophobic polymer, PEG, glycerin, agar and castor oil. Gingival plasters exhibited very good *in vivo* performance and appeared promising for gingival delivery of drugs, or for a local effect in the mouth.

Adhesive tablets consist of either monolithic, partially coated or multilayered matrices. Mucoadhesion is achieved mainly by cellulose or acrylic polymers, which offer almost immediate high adhesion performances for prolonged periods of time[45]. One of the first commercially available bioadhesive tablets (Aphtach[R]) was formulated by the group of Nagai[46] and intended for local treatment of aphtous stomatitis. The active ingredient (triamcinolone acetonide) was dispersed in the mixture of HPC and Carbopol 934, protected by an inert layer of lactose. Schor et al[47] have developed a nitroglycerine bioadhesive tablet (Susadrin[R]) for sublingual application, while Reckitt & Colman have produced an adhesive tablet, Buccastem[R], containing prochlorperazine for buccal use.

The limitations of adhesive tablets include the small surface area of contact, the lack of flexibility, the difficulty in obtaining a high release rate and possible irritation of mucosa. In order to overcome these problems, flexible adhesive films and laminated adhesive patches have been developed. Polymers used in mucoadhesive patches mainly include cellulose derivatives, natural gums and polycarylates. 3M Pharmaceuticals have developed a bioadhesive polymer patch formulation consisting of polyisobutylene, polyisoprene and Carbopol 934P[48]. Similarly, lignocaine, a local anaesthetic, has been formulated in the form of a mucoadhesive laminated patch to obtain soft tissue anaesthesia during dental procedures[49]. The effect of this "needle-free" application was comparable the one of infiltration anaesthesia.

Drug loaded adhesive films, based on bioadhesive polymers, represent another option in oral mucosal drug delivery. Kurosaki et al[50] reported the use of an HPC film for the delivery of propranolol. Rodu et al[51] prepared an adhesive film by complexing HPC with tannic and boric acid (Zilactin[R]) for the protection of oral ulcers against pain when drinking or eating.

Yamamoto et al[52] have described a mucoadhesive powder containing beclomethasone dipropionate and HPC. A significant increase in residence time compared to an oral solution was achieved. Mirth[53] has described three types of delivery devices for the sustained release of NaF: an aerosol consisting of guar gum and NaF-containing microcapsules, a small intraoral controlled-release pellet and a tablet with Carbopol 934 and HPMC. Scopps and Heiser[54] developed an intraoral adhesive bandage to protect mucosal wound after dental extraction, while Friedman and Steinberg[55] have described more advanced delivery devices that can be placed into the periodontal pocket and provide sustained release of tetracycline (hollow fibres) or chlorhexidine (thick slabs made of EC and HPC).

Semisolid systems are represented by adhesive ointments, usually containing poly(acrylic acid)[56,57] and adhesive gels. Ointments and gels are generally used for local therapy, since they can significantly prolong residence time and hence improve bioavailability. Gurny et al[58] studied the release of a local anaesthetic, febuverin, from a mixture of NaCMC and gelatin in polyethylene gel. Bremecker et al[59] investigated an adhesive gel of partly neutralised PMMA in water, loaded with tretinoin for the treatment of oral lishen planus; both *in vitro* and *in vivo* results were very encouraging.

6 RECENT WORK ON POLY(ACRYLIC ACID) SYSTEMS

6.1 The choice of bioadhesive polymers

Our research work has focused on the use of polymers based on poly(acrylic acid) (PAA). This group of synthetic polymers were found to possess very good mucoadhesive properties almost two decades ago and have attracted considerable interest in the field since that time. Indeed, there are many new PAA polymers on the market, some of them being intended specifically for bioadhesive formulations[60].

Poly(acrylic acid) polymer resins are long-chain, high-molecular weight, cross-linked molecules with a large number of COOH groups along the polymer backbone. This makes them hydrophilic, pH sensitive and capable of forming hydrogen bonds. In the dry state, the resin molecules are highly coiled and tightly packed. When placed in water, PAA polymers behave as anionic electrolytes. They dissociate and partially uncoil due to the repulsion of negative charges generated along the polymer chains. The subsequent swelling is caused and determined by the difference in osmotic pressure inside the vicinity of the polymer chains (cluster) and the bulk medium. In the presence of a neutralising agent, however, the processes of ionisation, uncoiling and hydration are largely enhanced, leading eventually to the formation of a stable three-dimensional polymer network. It is a usual procedure, therefore, to neutralise PAA resins when including them in the formulation of semisolid dosage forms.

There are two main groups of PAA homopolymers, produced by B.F.Goodrich and approved by the United States Pharmacopoeia (USP) and the National Formulary (NF):

- Carbomers (Carbopol^R resins), cross-linked with allylsucrose or allylpentaerythritol;
- Polycarbophils (Noveon^R resins), cross-linked with divinyl glycol.

The best known representative of the carbomer group is Carbopol 934 (and its purified version, Carbopol 934P, intended for oral and mucoadhesive applications). Several new carbomers have been recently introduced onto the market, namely Carbopol 974P (a "toxicologically preferred" alternative to 934P), its sodium salt Carbopol EX-214 and its less cross-linked variation Carbopol 971P[61]. Noveon AA-1 is the main representative of the polycarbophil group, having a tetra-functional cross-linker divinyl glycol, as opposite to the two-functional ones in carbomers.

6.2 The experimental techniques employed

We have examined the gel properties of PAA systems in an attempt to establish the link between the chemical structure of the polymers, the gelling properties and the bioadhesive properties. More specifically, we have assessed hydrogels formed from a range of neutralised PAA resins using a combination of two dynamic techniques, namely dielectric analysis and oscillatory rheology. In addition, an *in vitro* mucoadhesion tensile test was used in order to relate the structure of hydrogels to their adhesive performance.

Oscillatory rheology involves the application of an oscillatory shear stress to a sample and the subsequent measurement of the shear strain. As the measuring method is dynamic rather than static, many semisolid systems will respond differently as the frequency changes, i.e. the materials show viscoelasticity. In particular, at high frequencies gel systems may behave as elastic solids, whereby recovery is complete after removal of the applied stress. At low frequencies, however, the samples show predominantly viscous behaviour, whereby irreversible deformation occurs on application of the stress (i.e. the sample flows). At intermediate frequencies, the samples show components of both types of rheological behaviour. Analysis of data gives information on the structure of the system, particularly in terms of its rigidity and deformability.

The frequency dependent behaviour of viscoelastic systems may not be reliably expressed in terms of a single quantity, as it is necessary to state the elastic and viscous components separately. For dynamic methods, the best mathematical approach is shown to be the use of complex variables, with one component relating to the elastic behaviour and the other referring to the viscous behaviour. In oscillatory rheology, the energy stored and recovered per cycle of deformation is expressed in terms of a storage (elastic) modulus G', while the energy lost per cycle is referred to as a loss (viscous) modulus G'' [62]. The ratio of the two moduli is the tan δ value, given by tan $\delta = G''/G'$; hence it is a good indicator of the overall level of viscoelasticity at a particular frequency.

Dielectric analysis involves the application of an oscillatory electric field to a sample, again resulting in a response which is dependent on frequency. The two components of the dielectric response are most easily envisaged by considering the energy put into the system to be partially stored by processes such as dipole reorientation, and partially lost due to collisions caused by charge movement through the system. The energy stored is given by the capacitance C, while the energy lost as heat is given by the dielectric loss G/ω, where G is the conductance of the system. By studying both the absolute values and the relationship between these two components over a range of frequencies, it is possible to derive information on the structure of the sample.

The use of these two techniques in conjunction provides a novel approach to the characterisation of gel systems (or, indeed, to pharmaceutical materials in general); the dielectric analysis yields information on the movement of small molecules such as drugs through the system, while the results from oscillatory rheology relate to the structure of the polymer network. The two techniques therefore provide complementary information and allow a more profound insight into the structure and behaviour of the gel.

Bioadhesive performance of PAA hydrogels was studied using a Dynamic Contact Angle Analyser and a 30% w/w partially purified mucin gel, as a mucous substrate. Tensile tests were performed on a series of hydrogels, with a maximum detachment force taken as a mucoadhesive parameter. More details concerning the experimental procedures are given in the published literature[63].

6.3 Rheological characterisation

Rheological behaviour of PAA hydrogels is of considerable importance in terms of their general use in pharmaceutical preparations, but also because the rheological properties are known to be important contributors to mucoadhesive performance[2]. Indeed, evidence has recently been presented for mutual diffusion and interpenetration of polymer and mucin glycoprotein chains during the adhesion process[64], hence the molecular mobility, and therefore the rheological properties of these gels are of interest.

In our rheological studies[65], we have used a controlled-stress rheometer (Carri-med CSL 500,TA Instruments) and both continuous shear (flow) and oscillatory (dynamic) measurements, aiming to relate the rheological performance to the chemical structure of PAA polymers. Flow measurements were carried out by increasing the stress to 200Pa, holding and decreasing it back to 0, each stage taking 1 min. Figure 3[65] shows the flow curves of a range of sodium neutralised PAA hydrogels, whereby the lower values for shear rate under the same shear stress indicate a higher viscosity of the sample. It is evident that Carbopol 934 exhibits the highest viscosity and Carbopol EX-214 the lowest over the range of shear stresses under study. All the samples possess yield values, indicating that the gel networks exhibit a resistance to an external force before they start flowing (plastic behaviour). It is interesting to note that there are differences in the flow behaviour between Carbopol 974P and Carbopol EX-214, despite the fact that, after sodium neutralisation of 974P, the two gels should be identical in structure.

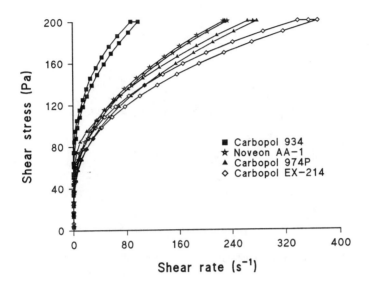

Figure 3. Flow curves of Na neutralised PA gel systems

Dynamic tests were performed over a three-decade frequency range (0.01 to 10Hz) in the linear viscoelastic region. The results of oscillatory measurements are expressed in terms of the tan δ (the ratio between loss and storage modulus) in Figure 4[65] . Carbopol 934 showed the lowest tan δ values over the whole frequency range, indicating the highest level of network elasticity. Carbopol 974P exhibited lower elastic properties, which reflect the structure differences between the two gel systems.

It is possible that for Carbopol 974P, a cross-linker other than allylsucrose has been used. Alternatively, it is known that Carbopol 934 is polymerised in benzene, Carbopol 974 and Noveon AA-1 in ethylacetate and Carbopol EX-214 in methanol[61]; hence the choice of solvent may be of relevance. Also, Carbopol 974P is treated with potassium during the manufacturing process[66], which may well be the source of difference, since monovalent cations generally lead to lower hydration of PAA polymer.

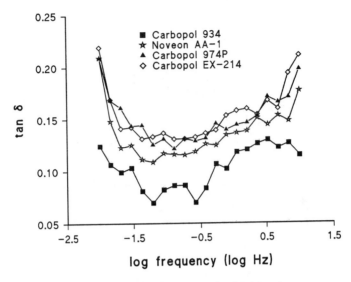

Figure 4. Tan δ values of Na neutralised PAA gel systems

Rheological differences observed between NaOH-neutralised Carbopol 974P and its sodium salt EX-214 may be explained by different degrees of ionisation between PAA and PAA-Na, respectively. In the case of Carbopol 974P, the addition of a strong base (NaOH) promotes dissociation of PAA, leading to the repulsion of like charges and the formation of an expanded gel network. In the case of Carbopol EX-214, the degree of ionisation is lower as the sodium salt is a weak electrolyte and, consequently, the hydration and the uncoiling of the molecular chains will be reduced, leading to a less elastic structure.

In another study[67] we have used a range of bases for neutralisation, these including one inorganic base (NaOH) and two organic amines: a primary amine tromethamine (TRIS) and a tertiary amine triethanolamine (TEA). Figure 5[67] shows the variation in storage moduli for carbomer (Carbopol 934P) and polycarbophil (Noveon AA-1) gel systems. The differences in elasticity between the two polymers are probably due to the type of cross-linking agent. Less pronounced variations in elastic moduli were caused by different neutralisers (Figure 5), but following a general trend that the base strength is inversely proportional to the extent of network elasticity. This is in good correlation with the finding of Lochhead et al[68], who suggested that a stronger base causes a smaller hydrodynamic volume (i.e. the extent of the uncoiling of polymer chains) of PAA gel.

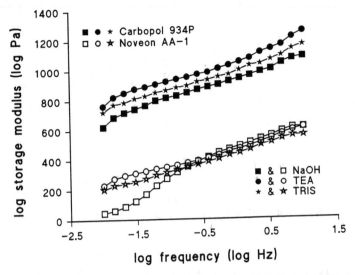

Figure 5. Variation of the storage modulus with frequency of Carbopol 934P and Noveon AA-1 gels, neutralised with a range of bases

6.4 Dielectric characterisation

In an investigation into the structure and properties of Carbopol 934 gels[69], we have explored the usefulness of low frequency dielectric spectroscopy in getting further information about the network structure. It was demonstrated that the dielectric response of Carbopol 934 over a frequency range of 10^4 to 10^2 Hz comprised two regions: the higher frequency region corresponding to a process dominated by the conductivity of the bulk liquid (i.e. the movement of charges through the gel network), while the lower frequency response corresponds to the establishment of a gel layer on the electrodes, through which charges may pass but with greater difficulty than through the bulk liquid. In general, therefore, dielectric analysis yields information on the movement of charges through the gel.

A further study[63] dealt with dielectric and other properties of a range of PAA gel systems, both unneutralised and neutralised with various bases. Typical dielectric spectra are shown in Figure 6[63] for Carbopol 974P with and without neutralisation with NaOH. The data indicate that the neutralised gels have a higher bulk conductivity than the unneutralised systems, either directly due to the addition of ionic base to the system or else due to the neutralising agent changing the ionisation state of the polymer chains while becoming integrated into the gel network itself.

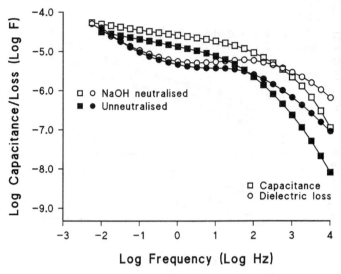

Figure 6. Low frequency dielectric response of Carbopol 974P gels, unneutralised and neutralised with NaOH

The logarithmic slope of the low frequency capacitance gives an indication of the integrity of the adsorbed layer, with a horizontal slope (zero) suggesting that the gel layer is acting as a perfect blocking layer through which charge may not penetrate, while a steeper (negative) slope indicated a "leaky" layer through which charge may pass. The low frequency capacitance slope is closer to the horizontal for the neutralised sample (Figure 6), revealing the greater charge blocking abilities (and hence physical integrity) of the gel layer. This is compatible with the conclusions drawn from the rheological studies, which demonstrated the greater rigidity of the neutralised systems.

In terms of different neutralising agents, the TEA neutralised samples showed greater low frequency conductivities than did the other systems (Figure 7[67]). This implies that for gels containing TEA there is an accumulation of charge in the gel layer located at the electrodes, even though there is no such trend seen for the high frequency (bulk) conductivity. Consequently, it may be suggested that the charges causing this effect are closely associated with polymer chains, rather than existing in a free state within the network. This effect corresponds very well with the highest elasticity obtained in the cases of all TEA neutralised PAA gel systems[63].

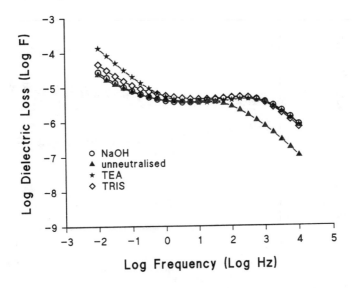

Figure 7. Low frequency dielectric loss of Noveon AA-1 gels, unneutralised and neutralised with a range of bases

6.5 The effects of drug addition

Chlorhexidine gluconate (CHG), a potent intra-oral antiseptic, has been used as a model drug in our studies on 2.5 and 5% Carbopol 934 gels[69,70]. It is known to have strong basic groups within the molecule, but, given the fact that the gel system already contains a significant quantity of base (8% of TEA), the addition of 0.1% CHG may be expected to have little effect on the gel structure. However, inspection of the rheological data (Figure 8[69]) indicates that at low frequencies, the presence of the drug has a profound effect on both the storage and loss moduli, with a peak being seen in the latter. At high frequencies, an increase in the storage modulus was seen on addition of the drug. It may be speculated that the addition of CHG may result in "thinning" the gel due to the screening of the anionic groups on the polymer chains, although the mechanism is likely to be a complex one.

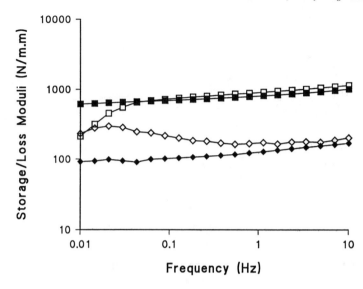

Figure 8. Effect of chlorhexidine gluconate on the rheological properties of Carbopol 934 (■, ♦: G′, G″ Carbopol 934. □, ◊: G′, G″ Carbopol 934 with 0.1% CHG)

The marked changes seen in the rheological data are mirrored by equally marked changes in the dielectric response, as shown in Figure 9[69]. The barrier layer is considerably less well defined in the presence of the CHG (seen by the higher negative slope of the low frequency capacitance), indicating that the barrier function of the gel layer has been profoundly disrupted. This correlates with the rheological data in that the gel structure has become considerably more open in the presence of CHG.

Both the dielectric and rheological profiles of the Carbopol 934 gels containing 5% propylene glycol were also altered on addition of a model drug at two concentration levels (0.05 and 0.10%)[69]. The effects on the dielectric response were not as marked as for the systems without propylene glycol, possibly due to the gel structure having already been to some extent disrupted by the propylene glycol itself. Moreover, there are indications of the presence of some synergistic effect between the drug and the propylene glycol. On the basis of the dielectric and rheological effects observed in this study, it should be stressed that it is not necessarily valid to correlate the dissolution behaviour of one drug with that of another, as the drug itself may alter the gel structure.

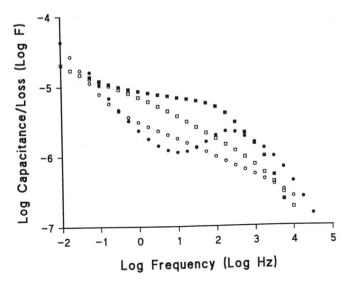

Figure 9. Effect of chlorhexidine gluconate on the dielectric properties of Carbopol 934 (■, •: C, G/ω Carbopol 934. □, ○: C, G/ω Carbopol 934 with 0.1% CHG)

6.6 Mucoadhesive performance and the effect of ageing

Maximum detachment forces for a range of freshly made PAA hydrogels obtained by an *in vitro* tensile method are presented in Table 1[71]. All the gel systems were 2.5% w/w, with the pH values between 6.8 and 7.2 for the neutralised samples and 3.1 to 3.2 for the unneutralised ones. In general, the mucoadhesive strength of Carbopol 934P and 974P systems is higher than Carbopol EX-214 and Noveon AA-1 samples (at least in the neutralised state). Similarly, the tan δ values are lower for the more mucoadhesive systems, indicating that mucoadhesion is greater in gels with the higher elastic component. The lower detachment force for Carbopol EX-214 systems compared with Carbopol 974P provides further evidence of the non-equivalence of these polymer gels, probably due to the differences in hydration, as previously discussed.

The significantly lower values obtained for the unneutralised gels do not support the assumption that adhesive gels of lower viscosity can make closer contact with the surface and so achieve better bioadhesion. In fact, the data obtained suggest that there is a relationship between viscoelastic properties of hydrogels and their bioadhesive performance. In particular, the tan δ (the ratio between viscous and elastic modulus) correlates with maximum detachment force in an inverse manner.

Table 1. Rheological and mucoadhesive parameters of the fresh and aged
PAA hydrogels (m ± sd)

Sample	Tan δ at 0.5 Hz (fresh sample)	Detach. force (mg) (fresh sample)	Tan δ at 0.5Hz (aged sample)	Detach. force (mg) (aged sample)
Carbopol 934P				
NaOH	0.134±0.001	3973±300	0.107±0.002	4341±143
TEA	0.125±0.001	4277±410	0.128±0.003	4200±252
TRIS	0.129±0.001	4290±342	0.130±0.002	4220±602
unneutr.	12.18±0.52*	1127±51	7.16±0.51*	1291±25
Carbopol 974P				
NaOH	0.200±0.012	4093±150	0.155±0.001	4368±218
TEA	0.159±0.001	4289±266	0.160±0.009	4300±320
TRIS	0.178±0.023	4226±149	0.155±0.002	4407±401
unneutr.	3.68±0.37*	1949±92	2.96±0.27*	1996±82
Noveon AA-1				
NaOH	0.286±0.045	3113±225	0.246±0.013	3252±251
TEA	0.236±0.005	3237±166	0.239±0.006	3184±302
TRIS	0.254±0.020	3336±183	0.276±0.018	3423±149
unneutr.	6.44±1.13*	1936±175	2.86±0.11*	1971±104
Carbopol EX-214	0.602±0.037	3291±124	0.530±0.015	3644±199

* Data taken at 1.129 Hz

The results obtained after a six-month ageing period at the room temperature provide further evidence for the existence of such a relationship (Table 1). For example, all NaOH neutralised samples have shown larger tan δ values after ageing, indicating an increase in overall elasticity compared with TEA and TRIS gels (which may be due to the fact that inorganic PAA salts possess slightly higher resistance to catalytic degradation than the organic ones[60]). Indeed, this trend corresponds very well with the increase in the maximum detachment force in Na hydrogels after ageing.

It is known that the viscosity of gels generally tends to decrease over time due to the elastic contraction of polymer molecules, so the falls in both elastic and storage modulus were expected. Figure 10[71] shows the changes in elastic moduli of a range of PAA hydrogels, before and after ageing. It is interested to note that the variations in G' between Carbopols 934P and 974P are more expressed after ageing than in the fresh state.

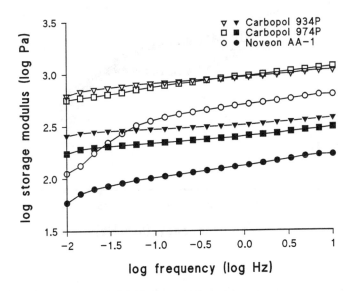

Figure 10. Variation in the storage modulus with frequency for a range of Na neutralised PAA gels, in the fresh (open symbols) and aged state (solid symbols)

Dielectric assessment of the fresh and aged PAA gel systems have revealed changes in both the low frequency (barrier layer) and high frequency (bulk) response. Typical spectra of the TEA neutralised gel system are given in Figure 11[71]. It is found that low frequency loss decreased on storage, while the high frequency response increased, which was valid for all the neutralised systems. This suggests that charge-carrying species have been released from the vicinity of the polymer into the bulk, either as a result of chemical degradation or changes in the neutralisation/dissociation equilibrium.

The proposed mechanism of charge redistribution inside the PAA network during storage is compatible with the observed decrease in G', as the release of neutralising ions from the bound state into the bulk aqueous phase may be expected to result in decrease in polymer chain ionisation and hence expansion, leading in turn to a lower elastic modulus. Dielectric analysis has therefore suggested an additional mechanism of ageing, i.e. a partial dissociation of the polyion-base salt over time.

Chemical Aspects of Drug Delivery Systems

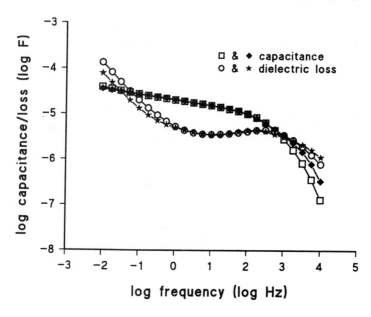

Figure 11. Low frequency dielectric response of fresh (open symbols) and aged (solid symbols) Noveon AA-1 gels, neutralised with TEA

6.7 Mucin/polymer mixtures

In order to optimise mucoadhesive dosage forms, it is helpful to have information on the structure of both mucoadhesive hydrogels and mucin/polymer mixtures. Moreover, the rheological behaviour of mucin/polymer mixtures could be used as an indicator of the biodhesive bond strength, as first suggested by Hassan and Gallo[72]. We have examined a series of mucin/PAA mixtures, using a partially purified pig gastric (PG) mucin as a mucus substitute. Rheological assessment comprised both frequency scans (from 0.01 to 10 Hz at 25°C) and temperature scans (from 10 to 90 and back to 10 °C, at the rate of 8 °C/min, and a constant frequency of 1 Hz).

Frequency sweep tests have revealed that the rheological synergism of the mixtures is a frequency-dependent phenomenon, determined by the structure of the polymer system[73]. Comparing the synergistic effect in terms of elastic modulus G', we have found that the level of cross-linking and the type of cross-linker of PAA resin strongly influence the rheological properties of a mucin/polymer mixture. Figure 12[73] shows that the least cross-linked polymer

Carbopol 971P exhibits rheological synergism over the whole frequency range, while Carbopol 974P and Noveon AA-1 have cross-points at certain frequencies. We believe that the time-dependent synergism of PAA/mucin mixtures could account for some unexpected findings by the other authors concerning the correlation between viscoelastic and adhesive parameters of PAA polymers.

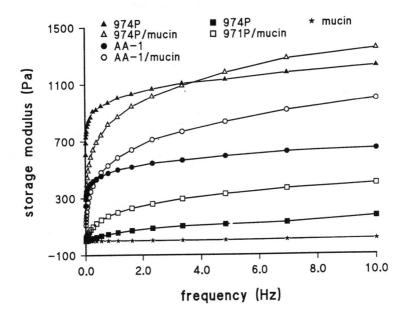

Figure 12. Variation of the storage modulus with frequency for mucin gel, TEA neutralised PAA gels and mucin/polymer mixtures

Temperature scans of the neutralised Carbopol 974P and 971P gels revealed the same trend of G' decrease upon heating, with much larger changes observed in the case of 974P[74]. This is likely to be caused by the differences in water distribution within the two networks: less cross-linked 971P has a "fishnet" gel structure whereby the water molecules are relatively uniformly distributed throughout the network, while 974P has a "fuzzball" gel structure[75]. The loss of some intra-cluster water and secondary chemical bonds, followed by the rearrangement of a polymer network, are the processes probably connected with the drop in elasticity in the case of Carbopol 974P gel system.

7 CONCLUSIONS

Bioadhesive dosage forms are a potentially highly useful means of delivering drugs to the body, perhaps particularly for topical or local administration where the mechanical trauma experienced by the dosage form may be minimised. In our work, we have attempted to emphasise the importance of understanding the gel properties of both the bioadhesive polymer and the mucin itself, as this represents an approach which has arguably not been adequately explored. Indeed, gels are notoriously difficult to study in depth; hence many important issues such as drug interactions with gels are not well understood. The relationship between the bioadhesion and the tan δ is of great interest and we are exploring this area in detail at present. In addition, the thermorheological studies seem to be extremely interesting, particularly for studying the interaction between the polymers and mucin. In this way, it is hoped that the physical characteristics of these systems may be considered alongside the biological features of bioadhesion.

8 REFERENCES

1. G. Hunt, P. Kearney and I.W. Kellaway in 'Drug Delivery Systems; Fundamentals and Techniques' P. Johnson and J.G. Lloyd-Jones eds. , Ellis Horwood Series in Medicine, Chichester, 1987, p. 180.

2. J.-M. Gu, J.R. Robinson and S.-H.,S. Leung, *CRC Crit. Rev. Therap. Drug Carr. Syst.*, 1988, **5**, 21.

3. R.B. Gandhi and J.R. Robinson, *Adv. Drug Del. Rev.*, 1994, **13**, 43.

4. R.W. Jeanloz in 'Glycoproteins, Their Composition, Structure and Function', A. Gottschalk ed., Elsevier, Amsterdam, 1972.

5. E.N. Chantler and P.R. Scudder, 'Ciba Foundation Symposium 109', Pitman, London, 1984, 180.

6. C. Robert and P. Buri, *Pharm. Acta Helv.*, 1986, **61**, 210.

7. M. Scawen and A. Allen, *Biochem. J.*, 1977, **163**, 363.

8. R. Kornfeld and S. Kornfeld, *Ann. Rev. Biochem.*, 1976, **45**, 217.

9. M.A. Longer and J.R. Robinson, *Pharm. Int.*, 1987, **7**, 114.

10. F.A. Meyer and A. Silberberg, 'Respiratory Tract Mucus', Ciba Foundation Symposium 54, Elsevier, Amsterdam, 1978, 203.

11. D. Duchêne, F. Touchard and N.A. Peppas, *Drug Dev. Ind. Pharm*, 1988, **24**, 283.

12. N.A. Peppas and P.A. Buri, *J. Control. Rel.*, 1985, **2**, 257.

13. G. Ponchel, F. Touchard, D. Duchêne and N.A. Peppas, *J. Control. Rel.*, 1987,**5**, 129-141.

14. M.J. Tobyn, J.P. Johnson and S.A.W. Gibson, *J. Pharm. Pharmacol.* (Suppl.), 1992, **44**, 1048.

15. S.A. Mortazavi, *Int. J. Pharm.*, 1995, **124**, 173.

16. C.-M. Lehr, J.A. Bouwstra, H.E. Boddé and H.E. Junginger, *Pharm. Res.*, 1992, **9**, 70.

17. M. Rillosi and G. Buckton, *Int. J. Pharm.*,1995, **117**, 75.

18. A.G. Mikos and N.A. Peppas, *Int. J. Pharm.*, 1989, **53**, 1.

19. S.A. Mortazavi and J.D. Smart, *J. Control. Rel.*, 1993, **25**, 197.

20. C.A. Serivener and C.W. Schantz, *J. Am. Dental Assoc.*, 1947, **35**, 644.

21. A.H. Kutscher, H.V. Zegarelli, F.E. Beube, N.W. Chiton, C. Berman, J.L. Mercadante, I.B. Stern and N. Roland, *Oral Surg. Oral Med. Oral Pathol.*, 1959, **12**, 1080.

22. J.L. Chen and G.N. Cyr, in 'Adhesion in Biological Systems', R.S. Manly ed., Academic Press, New York, 1970, 163.

23. R. Jimènez-Castellanos, H. Zia and C.T. Rhodes, *Drug Dev. Ind. Pharm.*, 1993, **19**, 143.

24. K. Park and J.R. Robinson, *Int. J. Pharm.*, 1984, **19**, 107.

25. J.D. Smart, I.W. Kellaway and H.E. Worthington, *J. Pharm. Pharmacol.*, 1984, **36**, 295.

26. K. Park and J.R. Robinson, *J. Control. Rel.*, 1985, **2**, 47.

27. R. Gurny, J.M. Meyer and N.A. Peppas, *Biomaterials*, 1984, **5**, 336.

28. M.E. de Vries and H.E. Boddé, *J. Biomed. Mater. Res.*, 1988, **22**, 1023.

29. M. Ishida, N. Nambu and T. Nagai, *Chem. Pharm. Bul.*, 1983, 1010.

30. S.A. Mortazavi and J.D. Smart, *Int. J. Pharm.*, 1995, **116**, 223.

31. S.A. Mortazavi and J.D. Smart, *J. Control. Rel.*, 1994, **31**, 207.

32. E.E. Hassan and J.M. Gallo, *Pharm. Res.*, 1990, 7; 491.

33. S. Rossi, M.C. Bonferoni, F. Ferrari, M. Bertoni and C Caramella, *Proced. 14th Pharn. Techn. Conf.,* Barcelona, April 1995, Vol. 2a, 464.

34. N.A. Peppas and A.G. Mikos, *S.T.P. Pharma,* 1989, **5**, 187.

35. H.L. Luessen, C.-M. Lehr, C.-O. Rentel, A.B.J. Noach, A.G. de Boer, J.C. Verhoef and H.E. Junginger, *J. Control. Rel.,* 1994, **29**, 329.

36. M.A. Longer, H.S. Ch'ng and J.R. Robinson, *J. Pharm. Sci.,* 1985, **74**, 406.

37. R. Khosla and S.S. Davis, *J. Pharm. Pharmacol.,* 1987, **39**, 47.

38. R. Ito, Y. Machida, T Sannan and T. Nagai, *Int. J. Pharm.,* 1990, **61**, 109.

39. Müller and Heinemann, in 'Bioadhesion - Possibilities and Future Trends', R. Gurny and H.E. Junginger eds., Wissenschaftliche Verlagsgesselschaft, Stuttgart, 1990, p. 202-213.

40. M.E. de Vries, H.E. Boddé, H.I. Busscher and H.E. Junginger, *J. Biomed. Mater. Res.,* 1988, **22**, 1023-1032.

41. M.E. de Vries, H.E. Boddé, J.C. Verhoef and H.E. Junginger, *Crit. Rev. Therap. Drug Carr. Systems,* 1991, **8**, 271.

42. H..Merkle. R Anders, A. Wermerskirchen, S.C. Raehs and G. Wolany, in 'Peptide and Protein Drug Delivery', V.H.L.Lee ed., Plenum Press, New York, 1992.

43. J.D. Smart, *Advanced Drug Delivery Reviews,* 1993, **11**, 253-270.

44. T. Nagai, *Med. Res. Rev.,* 1986, **6**, 227-242.

45. G. Ponchel, *Advanced Drug Delivery Reviews,* 1994, **13**, 75-87.

46. T. Nagai and Y. Machida, *Pharm. Int.,* 1985, **6**, 196.

47. J.M. Schor, S.S. Davis, A Nigalaye and S. Bolton, *Drug Dev. Ind. Pharm.,* 1983, **9**, 1359.

48. J.-H. Guo, *J. Control. Rel.,* 1994, **28**, 272.

49. I.M. Brook, G.T. Tucker, E.C. Tuckley and R.N. Boyes, *J. Control. Rel.,* 1989, **10**, 183.

50. Y. Kurosaki, T Takatori, M. Kitayama, T. Nakayama and T Kimura, *J. Pharmacobio. Dyn.,* 1988, **11**, 824-832.

51. B. Rodu, C.M. Russell and A.J. Desmarais, *J. Oral Pathol.,* 1988, **17**, 564-567.

52. M. Yamamoto, K. Okabe, J. Kubo, T. Naruchi, H. Ikura, Y. Suzuki and T. Nagai, *STP Pharma,* 1989, **5**, 878-885.

53. D.B. Mirth, *Pharmacol. Ther. Dent.,* 1980, **5**, 59.

54. I.W. Scopps and R.A. Heiser, *J. Biomed. Materials Res.*, 1967, **1**, 371.

55. M. Friedman and D. Steinberg, *Pharm. Res.*, 1990, **7**, 313.

56. M. Ishida, N. Nambu and T. Nagai, *Chem. Pharm. Bull.*, 1982, **30**, 980-983.

57. M. Ishida, N. Nambu and T. Nagai, *Chem. Pharm. Bull.*, 1983, **31**, 4561-4564.

58. R. Gurny, J. Meyer and N.A. Peppas, *Biomaterials*, 1984, **5**, 336.

59. K.-D. Bremecker, H. Strempel and G. Klein, *J. Pharm. Sci.*, 1984, **73**, 548-552.

60. B..Goodrich Speciality Chemicals, 'Polymers for pharmaceutical applications. I. General overview', s.a.

61. B.F.Goodrich Speciality Chemicals, 'Carbopol resins handbook', 1992.

62. J.D. Ferry, 'Viscoelastic properties of polymers', Wiley, New York, 1970.

63. S. Tamburic and D.Q.M. Craig, *J. Control. Rel.*, 1995 (in press)

64. E. Jabbari, N. Wisniewski and N.A. Peppas, *J. Control. Rel.*, 1993, **26**, 99.

65. S. Tamburic and D.Q.M. Craig, *Pharm. Sci.*, 1995, **1**, 107.

66. M.J. Durrani, A. Andrews, R. Whitaker and S.C. Benner, *Drug Dev. Ind. Pharm.*, 1994, **20**, 2439.

67. S. Tamburic and D.Q.M. Craig, *14th Pharm. Technol. Confer.*, 1995, **2**, 188.

68. R.Y. Lochhead, A.C. Eachus and K.-D. Bremecker, in 'Kosmetikjährbuch 1992' B. Ziolkowsky ed. , Verlag für chemische Industrie, H. Ziolkowsky KG, Augsburg, 1992, p. 69.

69. D.Q.M. Craig, S. Tamburic, G. Buckton and J.M. Newton, *J. Control. Rel.*, 1994, **30**, 213.

70. G. Buckton and S. Tamburic, *J. Control. Rel.*, 1992, **20**, 29.

71. S. Tamburic and D.Q.M. Craig, *Pharm. Res.*, 1996 (in press)

72. E.E. Hassan and J.M. Gallo, *Pharm. Res.*, 1990, **7**, 491.

73. S. Tamburic and D.Q.M. Craig, unpublished data

74. S. Tamburic and D.Q.M. Craig, *Pharm. Res.*, Suppl., 1995 (in press)

75. B.F. Goodrich Speciality Chemicals, 'Carbopol 971P for pharmaceutical applications', Technical note CP4, 1994

ACKNOWLEDGEMENT

We would like to thank the Wellcome Trust for their kind provision of a Travelling Research Fellowship (grant number 040086/Z/93/Z) for Dr Tamburic.

Some Novel Aspects of Transdermal Drug Delivery

Kenneth A. Walters

AN-EX, REDWOOD BUILDING, KING EDWARD VII AVENUE, CARDIFF CF1 3XF, UK

1 INTRODUCTION

The increasing importance of the skin as a route for systemic drug delivery has generated a considerable amount of data which, collectively, has resulted in a deeper understanding of the mechanisms of transport across the skin and the process of percutaneous absorption[1-3]. Because of the inherent barrier characteristics of the skin many studies have concentrated on increasing the rate of drug delivery into and through the stratum corneum. Methodologies involved have ranged from the investigation of chemical penetration enhancers[4-6] to the use of physical forces such as iontophoresis and ultrasound[7,8], all with various degrees of success or failure. While these permeation enhancement technologies are very important to the continued evolution of systemic transdermal therapy, they introduce a complication in a fundamental requirement of such products, adhesion to the skin. This paper reviews the available chemical skin penetration enhancement technologies for improving transdermal delivery for systemic therapy and discusses their impact on adhesive properties and development.

2 THE SKIN BARRIER AND PENETRATION ENHANCEMENT

2.1 The Barrier

The permeability properties of human skin have been reasonably well characterised and have been extensively reviewed elsewhere[9,10]. There has been a considerable historical debate concerning routes of penetration across the skin and the relative importance of appendages such as sweat ducts and hair follicles[11,12]. It has, however, become apparent that the major barrier function of skin is provided by the stratum corneum, a coherent membrane comprised of morphologically distinct protein and lipid domains. The majority of the protein of the stratum corneum is composed of intracellular keratin filaments which are cross-linked by inter-molecular disulphide bridges[13,14]. Throughout the epidermis cell to cell contact is maintained via desmosomes and cadherins[15] but these become degraded, possibly with the help of the stratum corneum chymotryptic enzyme[16], as the keratinocytes approach the surface of the skin and desquamation occurs.

The lipid domain comprises an organised distribution of intercellular lamellae derived from intracellular granules secreted during the epithelial differentiation process[17]. Although there is little doubt that the protein domain, in particular the cornified envelope of the keratinocyte, can play a significant role in the barrier and protective function of the stratum corneum[18,19], the collective evidence strongly suggests that the intercellular lipid lamellae constitute the major barrier to diffusion.

The composition of the stratum corneum intercellular lipids is unique in biological systems (Table 1). A remarkable feature is the lack of phospholipids and preponderance of cholesterol and ceramides.

Table 1   *Composition of Human Stratum Corneum Lipids*

Lipid	% (w/w)	mol %
Cholesterol esters	10.0	7.5[a]
Cholesterol	26.9	33.4
Cholesterol sulphate	1.9	2.0
Total cholesterol derivatives	38.8	42.9
Ceramide 1	3.2	1.6
Ceramide 2	8.9	6.6
Ceramide 3	4.9	3.5
Ceramide 4	6.1	4.2
Ceramide 5	5.7	5.0
Ceramide 6	12.3	8.6
Total ceramides	41.1	29.5
Fatty acids	9.1	17.0[a]
Others	11.1	10.6[b]

[a] Based on C16 alkyl chain. [b] Based on MW of 500. Source ref. 20.

Six classes of ceramides (designated ceramides 1 to 6) have been isolated and identified in human stratum corneum[21]. Their structures contain long-chain bases, such as sphinganine and 4-hydroxysphinganine, N-acetylated by different fatty acids. Because of its unique structure, ceramide 1, an acylceramide, may function as a stabiliser of the intercellular lipid lamellae. This molecule consists of the usual sphingosine base with an amide-linked ω-hydroxyacid and a nonhydroxyacid ester linked to the ω-hydroxyl group[20]. It is possible that ceramide 1 acts as a "molecular rivet" in the intercellular lipid lamellae of the stratum corneum. There is also strong evidence to indicate that the intercellular lipid lamellae are further stabilised by chemical bonds between the long-chain ceramides and glutamate residues on the corneocyte protein envelope[19,22].

Overall, the intercellular lipid lamellae appear to be highly structured, very stable and constitute a highly effective barrier to chemical penetration and permeation. In order to increase the number of candidate drugs for systemic transdermal therapy it will be necessary to enhance drug delivery either by formulation strategies or by modifying the skin permeability barrier.

2.2 Penetration Enhancement

2.2.1 Formulation Strategies. As discussed above, the stratum corneum presents a considerable lipophilic barrier to the permeation of chemicals. On the other hand, the viable epidermis, which is immediately below the stratum corneum, is hydrophilic in nature and can act as a rate-limiting step to the absorption of highly lipophilic drugs. The overall result is that the optimal characteristic for percutaneous absorption is that the permeant be reasonably soluble in both hydrophilic and hydrophobic media[23] and, as such, is intimately linked to the oil-water partition coefficient of the permeant. It is well established that a principal driving force for diffusion across the skin is the thermodynamic activity of the permeant in the donor vehicle. This activity is reflected by the concentration of the permeant in the donor vehicle as a function of its saturation solubility within that medium. The closer to saturation concentration the higher is the thermodynamic activity and the greater is the escaping tendency of the permeant from the vehicle. This principle has been extensively and successfully utilized in pharmaceutical formulations in attempts to enhance percutaneous absorption of drugs[24].

The realisation that the thermodynamic activity of the drug in the formulation was a significant driving force for the release and penetration of the active into the skin, and that this characteristic could be manipulated, gave rise to the hypothesis that skin penetration may be enhanced if the thermodynamic activity was increased above unity (i.e., by the creation of a supersaturated formulation). The physical chemistry of supersaturated solutions is beyond the scope of this chapter and the reader is referred to the excellent review of Davis and Hadgraft[25]. Briefly, supersaturated systems can be created using a binary mixture in which one component is a good solvent for the solute and the other component is a non-solvent. Slow addition of the non-solvent to the solvent (both presaturated with solute) creates the supersaturated solution. For example, the degree of supersaturation attainable for hydrocortisone acetate in binary propylene glycol-water cosolvent systems can be as high as eight-fold saturation levels (Table 2). Because of the high potential for crystal growth in these systems it is necessary to add an antinucleant such as hydroxypropylmethyl cellulose. Increased transport across polydimethylsiloxane membranes has been shown to correlate with degree of supersaturation for hydrocortisone acetate[26] and piroxicam[27]. In addition, supersaturated formulations of hydrocortisone acetate have been shown to be bioequivalent to regular formulations containing much higher concentrations of the active[28].

Table 2   *Hydrocortisone Acetate Supersaturation in Propylene Glycol-Water Mixed Solvents*

Solvent A[a]	Solvent B[b]	Hydrocortisone Acetate % (w/w)	Degree of Supersaturation[c]
1	0	0.080	0.9
1	1	0.040	4.0
1	2	0.027	6.8
1	3	0.020	8.0
1	7	0.010	7.0
1	15	0.005	4.0

[a] 0.08% (w/w) hydrocortisone acetate in 90:10 propylene glycol:water.
[b] Water plus antinucleant. [c] Saturated = 1.0.

Supersaturated transdermal delivery devices have also been described although, in these cases, stability considerations limit the degree to which the systems can be supersaturated. Supersaturated systems containing nitroglycerine or isosorbide dinitrate have been formed in polymer films[29]. Films containing two-fold saturated drug are claimed to be stable on long-term storage. Similarly, Jenkins[30] describes the development of saturated and supersaturated transdermal drug-in-adhesive systems and demonstrates their use with norethisterone and œstradiol. In the preparation of these systems the active agent is dissolved in a mixture of solvents, at least one of which has a boiling point above that of the solvent used as a vehicle for the adhesive. The degree of saturation or supersaturation is dictated by the selected solvent mix and, because there are a large number of suitable solvents, the system is suitable for a wide range of drugs. Also embodied in the patent application is the suggestion that at least one of the remaining solvents may act as a skin permeation enhancer and in this respect propylene glycol-diethyltoluamide and n-methyl-pyrrolidone-diethyltoluamide are included in the preferred solvent systems. It would be interesting to evaluate the potential of developing systems in which both the drug and the permeation enhancer were in the supersaturated state.

2.2.2 *Penetration Enhancers.* Chemical penetration and permeation enhancers comprise a diverse group of compounds which are both natural (e.g. terpenes[31]) and synthetic (e.g. surfactants[32,33]). Included in a long list of compounds that have been shown to act as skin penetration enhancers are water[34], organic solvents[35,36], phospholipids[37], simple alkyl esters[38], long-chain alkyl esters[39], fatty acids[40], urea and its derivatives[41,42] and pyrrolidones[43]. Synergy between penetration enhancers and the solvents in which they are applied is often observed[44]. Thus, for example, the sesquiterpene α-bisabolol is much more effective in enhancing the skin penetration of 5-fluorouracil when applied in combination with propylene glycol[45].

In some cases molecules with specific potential as skin penetration enhancers have been designed and synthesised. Of the latter, 2-n-nonyl-1,3-dioxolane (SEPA®)[46-48], 1-dodecylazacycloheptan-2-one (Azone®)[49], 1-[2-(decylthio)ethyl]azacyclopentan-2-one (HPE-101)[50-52] and 4-decyloxazolidin-2-one (Dermac™ SR-38)[53] have all been shown to possess enhancing properties on the skin penetration and permeation of a variety of permeants. Many derivatives of Azone have been synthesised and evaluated for skin penetration enhancement activity[54-57]. Thus, for example, Hoogstraate et al[54] demonstrated that several linear alkyl chain analogues of Azone were less effective in enhancing the human skin penetration rate of desglycinamide arginine vasopressin than the parent molecule (Figure 1).

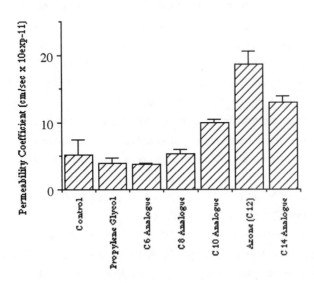

Figure 1 *Influence of Azone and analogues on the permeability coefficients of desglycinamide arginine vasopressin through human stratum corneum.*

Although Azone has been shown to be a useful skin penetration enhancer for many diverse compounds, its use is not always associated with success. For example, Baker and Hadgraft[58] have evaluated the use of Azone in vehicles containing the antiviral agent 4[-(2-chloro-4-methoxy)phenoxy] hexyl-3,5-heptanedione (Arildone) and found that penetration through human skin was not increased above levels found with a propylene glycol vehicle. The most likely reason for the lack of Azone skin permeation enhancement for Arildone was thought to be the highly lipophilic nature of the drug which resulted in poor partitioning from the stratum corneum into the viable epidermis. Azone enhancement activity is believed to occur in the stratum corneum intercellular lipid lamellae. Since a similar mechanism of action is proposed for HPE-101[51] it follows that the skin permeation of highly lipophilic compounds would also not be enhanced using this compound alone. Indeed, the selection of a suitable co-solvent for HPE-101 appears critical and, in some cases, it is necessary to combine the enhancer with agents such as cyclodextrins to improve activity[52].

On the other hand, SEPA® (2-n-nonyl-1,3-dioxolane; US Patent 4,861,764) has been shown to be a much more versatile penetration enhancer in terms of its ease of formulation, chemical stability and its ability to enhance the skin penetration of a wide variety of compounds of varying physicochemical characteristics[48,59-61]. Permeants evaluated include indomethacin, ibuprofen, minoxidil, acyclovir, caffeine, econazole, papaverine, progesterone, œstradiol, hexyl nicotinate and a 1200 dalton polypeptide. The degree of skin penetration enhancement using SEPA is dependent on the physicochemical characteristics of the permeant. For example, following application of indomethacin in a simple ethanol-propylene glycol vehicle to human skin in vitro, cumulative absorption over 24 hours amounted to 0.7% of the applied dose. The addition of 2% SEPA to the vehicle increased the 24 hour absorption value to 23% of the applied dose[62]. Furthermore, in comparative studies between SEPA and Azone, SEPA has been shown to be a more effective human skin permeation enhancer for indomethacin[60] (Figure 2).

More recently, SEPA has been shown to enhance the penetration of econazole into human skin[47] in vitro. Levels of econazole, in both the epidermis (stratum corneum and viable epidermis) and dermis, were significantly increased when SEPA was incorporated into gel vehicles at 5% and 10%. Thus, following application of econazole in a 5% SEPA gel, approximately 12% of the applied dose was recovered from the epidermis following 24 hours exposure. Interestingly the 5% SEPA gel was more effective than the 10% SEPA gel suggesting a concentration dependency in SEPA activity (Figure 3). The concentration dependency may be the result of a saturable phenomenon within the skin[63] or a decrease in the thermodynamic activity of econazole in the SEPA gels as a consequence of increased solubility of the permeant in the vehicle. In Figure 3 the data obtained using SEPA based gels is compared with that obtained using a control gel (containing no SEPA) and a commercial cream containing econazole nitrate.

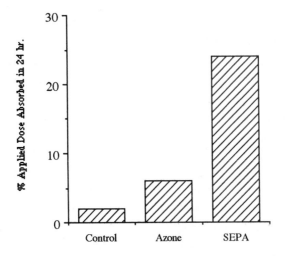

Figure 2 *Influence of SEPA and Azone on the permeation of indomethacin through human skin.*

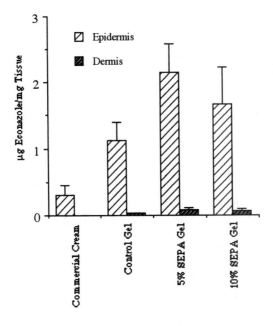

Figure 3 *Influence of SEPA on the human skin uptake of econazole.*

In addition to its effectiveness as a skin penetration enhancer, SEPA has been evaluated in a large number of preclinical and clinical safety studies[64]. The results of these studies show that SEPA does not possess dermal or systemic toxicological effects and there is no evidence of delayed allergic reactions or photosensitisation. In summary, the skin penetration enhancement properties of SEPA, together with extensive data demonstrating safety and ease of incorporation into a variety of dermatological and transdermal formulations, suggests that SEPA will be a very useful tool for the formulator.

3 PENETRATION ENHANCERS AND PRESSURE SENSITIVE ADHESIVES

The need for chemical permeation enhancers to be incorporated into transdermal delivery systems has has been described above. However, their use has been limited by lack of compatibility with the available pressure sensitive adhesives. Materials used as adhesives in transdermal systems are presently limited to polyisobutylenes, polyacrylates and silicones. The requirements of pressure sensitive adhesives for use in transdermal systems is daunting. Apart from the standard characteristics the adhesive must possess good biocompatibility (low irritancy, nonsensitiser, low acute and chronic toxicity); have good adhesion to oily, wet, wrinkled and hairy skin; and be resistant but permeable to water[65]. The standard evaluation of pressure sensitive adhesives is based on three fundamental properties: peel adhesion, tack and shear strength[66]. All three properties can be influenced by the addition of skin penetration enhancers. In addition to these general properties, adhesives designed for application to the skin are required to release the drug (and, of course, the enhancer) and should leave no residues on removal from the skin[66,67]. Drug release from silicone adhesives can be modified using copolymers. For example, hydrophilic drug release rates from polydimethylsiloxane matrices can be increased by up to three orders of magnitude using polydimethylsiloxane/polyethylene oxide graft copolymers[68].

On incorporation of a skin permeation enhancer, however, there may be an alteration in drug release rates and adhesive properties. For example, addition of 1% urea to a polyacrylate type pressure sensitive adhesive results in a loss of adhesion and skin contact cannot be maintained over the required period[67]. An obvious method to reduce the influence of both drug and enhancer on adhesive properties is to design the transdermal system such that there is no contact between these constituents and the adhesive[69]. In this case, the adhesive is present in a boundary laminate which surrounds the drug/enhancer releasing layer. One disadvantage of this type of system, however, is that the drug/enhancer releasing layer may not remain in intimate contact with the skin.

If high levels of liquid skin penetration enhancers are incorporated into matrix type drug-in-adhesive transdermal patches, there is likely to be a loss in cohesiveness. This will result in patch slipping and skin residues following patch removal. Cohesive strength, however, can be increased by high levels of cross linking in acrylate adhesives but this may alter both long-term bonding and drug release rates. These problems may be overcome by the use of grafted copolymer adhesives such as ARcare® ETA Adhesive Systems[70]. In this system reinforcement is achieved mainly through phase separation of the side chain within the continuous polymer network. A variety of side chains are available and enhancer concentrations up to 30% can be incorporated without seriously affecting its adhesive properties. This work has been limited to fatty acid ester type enhancers and usefulness with other enhancer types remains to be established. It may also be possible to maintain adhesive properties in the presence of skin penetration enhancers by using blends of acrylic copolymers with different molecular weights[71,72].

It is important to appreciate that it is also a fundamental requirement that the enhancer, as well as the drug, is released by the adhesive. Furthermore, it is probable that the presence of the enhancer may increase the skin permeation of other formulation excipients and that this may have an impact on local toxicity. Much remains to be done in the field of enhancer incorporation into transdermal drug delivery systems and it is encouraging to observe the increasing efforts of adhesive manufacturers in this sphere.

References

1. B.W. Barry, 'Dermatological Formulations, Percutaneous Absorption', Marcel Dekker, New York, 1983.
2. J. Hadgraft and R.H. Guy (eds.), 'Transdermal Drug Delivery', Marcel Dekker, New York, 1989.
3. V.P. Shah and H.I. Maibach (eds.), 'Topical Drug Bioavailability, Bioequivalence, and Penetration', Plenum Press, New York, 1993.
4. K.A. Walters and J. Hadgraft (eds.), 'Pharmaceutical Skin Penetration Enhancement', Marcel Dekker, New York, 1993.
5. J. Hadgraft and K.A. Walters in 'Drug Absorption Enhancement', A.G. de Boer (ed.), Harwood Academic Publishers, Amsterdam, 1994, Chapter 6, p. 177.
6. E.W. Smith and H.I. Maibach (eds.), 'Percutaneous Penetration Enhancers', CRC Press, Boca Raton, 1995.
7. M.B. Delgado-Charro and R.H. Guy, *Internat. J. Pharmaceut.*, 1995, **117**, 165.
8. J-P. Simonin, *J. Contr. Rel.*, 1995, **33**, 125.
9. M.S. Roberts in 'Prediction of Percutaneous Penetration', R.C. Scott, R.H. Guy, J. Hadgraft and H.E. Boddé (eds.), IBC Technical Services, London, 1991, Chapter 21, p. 210.
10. R.C. Wester and H.I. Maibach, *Clin. Pharmacokin.*, 1992, **23**, 253.
11. A.C. Lauer, L.M. Lieb, C. Ramachandran, G.L. Flynn and N.D. Weiner, *Pharm. Res.*, 1995, **12**, 179.
12. F. Hueber, H. Schaefer and J. Wepierre, *Skin Pharmacol.*, 1994, **7**, 237.
13. T-S. Sun and H. Green, *J. Biol. Chem.*, 1978, **253**, 2053.
14. P.M. Steinert, A.C.T. North and D.A.D. Parry, *J. Invest. Dermatol.*, 1994, **103**, 19S.
15. M. Amagai, *J. Invest. Dermatol.*, 1995, **104**, 146.
16. B. Sondell, L-E. Thornell and T. Egelrud, *J. Invest. Dermatol.*, 1995, **104**, 819.
17. P.W. Wertz and D.T. Downing, *Science*, 1982, **217**, 1261.

18. M.E. Ming, H.A. Daryanani, L.P. Roberts, H.P. Baden and J.C. Kvedar, *J. Invest. Dermatol.*, 1994, **103**, 780.
19. N.D. Lazo, J.G. Meine and D.T. Downing, *J. Invest. Dermatol.*, 1995, **105**, 296.
20. P.W. Wertz and D.T. Downing in 'Transdermal Drug Delivery', J. Hadgraft and R.H. Guy (eds.), Marcel Dekker, New York, 1989, Chapter 1, p. 1.
21. W.P Wertz, M.C. Miethke, S.A. Long, J.S. Strauss and D.T. Downing, *J. Invest. Dermatol.*, 1985, **84**, 410.
22. D.T. Downing, *J. Lipid Res.*, 1992, **33**, 301.
23. C. Surber, K-P. Wilhelm and H.I. Maibach, *Eur. J. Pharm. Biopharm.*, 1993, **39**, 244.
24. J. Ostrenga, J. Haleblian, B. Poulsen, B. Ferrell, N. Mueller and S. Shastri, *J. Invest. Dermatol.*, 1971, **56**, 392.
25. A.F. Davis and J. Hadgraft in 'Pharmaceutical Skin Penetration Enhancement' K.A. Walters and J. Hadgraft (eds.), Marcel Dekker, New York, 1993, Chapter 11, p. 243.
26. A.F. Davis and J. Hadgraft, *Int. J. Pharmaceut.*, 1991, **76**, 1.
27. M.A. Pellett, A.F. Davis and J. Hadgraft, *Int. J. Pharmaceut.*, 1994, **111**, 1.
28. R. Marks, P.J. Dykes, J. Gordon, G. Hanlan and A.F. Davis, British Society of Investigative Dermatology Annual Meeting. Sheffield, 1992.
29. M. Dittgen and R. Bombor, German Patent DD 217989 A1.
30. A.W. Jenkins, UK Patent Application GB 2249956 A.
31. B.W. Barry and A.C. Williams, in 'Pharmaceutical Skin Penetration Enhancement' K.A. Walters and J. Hadgraft (eds.), Marcel Dekker, New York, 1993, Chapter 4, p. 95.
32. E.J. French, C.W. Pouton and K.A. Walters, in 'Pharmaceutical Skin Penetration Enhancement' K.A. Walters and J. Hadgraft (eds.), Marcel Dekker, New York, 1993, Chapter 5, p. 113.
33. P.P. Sarpotdar and J.L. Zatz, *Drug Dev. Ind. Pharm.*, 1986, **12**, 1625.
34. M.S. Roberts and M. Walker, in 'Pharmaceutical Skin Penetration Enhancement' K.A. Walters and J. Hadgraft (eds.), Marcel Dekker, New York, 1993, Chapter 1, p. 1.
35. B. Berner, G.C. Mazzenga, J.H. Otte, R.J. Steffens, R-H. Juang and C.D. Ebert, *J. Pharm. Sci.*, 1989, **78**, 402.
36. T.J. Franz, in 'Percutaneous Penetration Enhancers' E.W. Smith and H.I. Maibach (eds.), CRC Press, Boca Raton, 1995.
37. G.P. Martin, in 'Pharmaceutical Skin Penetration Enhancement' K.A. Walters and J. Hadgraft (eds.), Marcel Dekker, New York, 1993, Chapter 3, p. 57.
38. D.R. Friend and J. Heller, in 'Pharmaceutical Skin Penetration Enhancement' K.A. Walters and J. Hadgraft (eds.), Marcel Dekker, New York, 1993, Chapter 2, p. 31.
39. M. Dohi, F. Kaiho, A. Suzuki, N. Sekiguchi, N. Nakajima, H. Nomura and Y. Kato, *Chem. Pharm. Bull.*, 1990, **38**, 2877.
40. M.A. Yamane, A.C. Williams and B.W. Barry, *Int. J. Pharmaceut.*, 1995, **116**, 237.
41. C.K. Kim, J-J. Kim, S-C. Chi and C.K. Shim, *Int. J. Pharmaceut.*, 1993, **99**, 109.
42. A.C. Williams, in 'Percutaneous Penetration Enhancers' E.W. Smith and H.I. Maibach (eds.), CRC Press, Boca Raton, 1995.
43. S.A. Akhter and B.W. Barry, *J. Pharm. Pharmacol.*, 1984, **37**, 27.
44. B. Møllgaard, in 'Pharmaceutical Skin Penetration Enhancement' K.A. Walters and J. Hadgraft (eds.), Marcel Dekker, New York, 1993, Chapter 10, p. 229.
45. R. Kadir and B.W. Barry, *Int. J. Pharmaceut.*, 1991, **70**, 87.
46. O. Doucet, H. Hagar and J-P. Marty, *STP. Pharm. Sci.*, 1991, **1**, 89.
47. R. Gyurik, S. Krauser, E. Gauthier and C. Reppucci, *Prediction of Percutaneous Penetration*, 1995, **4a**, 36.
48. R.W. Pelham and C.M. Samour, *Proceed. Intern. Symp. Control. Rel. Bioact. Mater.*, 1995, **22**, 694.

49. J. Hadgraft, D.G. Williams and G. Allan, in 'Pharmaceutical Skin Penetration Enhancement' K.A. Walters and J. Hadgraft (eds.), Marcel Dekker, New York, 1993, Chapter 7, p. 175.
50. T. Yano, N. Higo, K. Furukawa, M. Tsuji, K. Noda and M. Otagiri, *J. Pharmacobiodyn.*, 1992, **15**, 529.
51. T. Yano, N. Higo, K. Fukuda, M. Tsuji, K. Noda and M. Otagiri, *J. Pharm. Pharmacol.*, 1993, **45**, 775.
52. H. Adachi, T. Irie, K. Uekama, T. Manako, T. Yano and M. Saita, *Eur. J. Pharm. Sci.*, 1993, **1**, 117.
53. W.R. Pfister and V.J. Rajadhyaksha, *Pharm. Res.*, 1995, **12**, S-280.
54. A.J. Hoogstraate, J. Verhoef, J. Brussee, A.P. IJzerman, F. Spies and H. Boddé, *Int. J. Pharmaceut.*, 1991, **76**, 37.
55. B.B. Michniak, M.R. Player, J.M. Chapman and J.W. Sowell, *Int. J. Pharmaceut.*, 1993, **91**, 85.
56. B.B. Michniak, M.R. Player, L.C. Fuhrman, C.A. Christensen, J.M. Chapman and J.W. Sowell, *Int. J. Pharmaceut.*, 1994, **110**, 231.
57. B.B. Michniak, M.R. Player, D.A. Godwin, C.A. Phillips and J.W. Sowell, *Int. J. Pharmaceut.*, 1995, **116**, 201.
58. E.J. Baker and J. Hadgraft, *Pharmaceut. Res.*, 1995, **12**, 993.
59. C.M. Samour, MacroChem Corporation, personal communication, 1995.
60. C.M. Samour, L.G. Donaruma, S. Daskalakis, B.S. Fulton, J-P. Marty, A.M. Dervault, J.F. Chanez and O. Doucet, *Proceed. Intern. Symp. Control. Rel. Bioact. Mater.*, 1989, **16**, 183.
61. A.R. Diani, K.L. Shull, M.J. Zaya and M.N. Brunden, *Skin Pharmacol.*, 1995, **8**, 221.
62. J-P. Marty, A.M. Dervault, J.F. Chanez, O. Doucet and C.M. Samour, *Proceed. Intern. Symp. Control. Rel. Bioact. Mater.*, 1989, **16**, 179.
63. M. Hori, K-C. Moon, H.I. Maibach and R.H. Guy, *J. Contr. Rel.*, 1991, **16**, 263.
64. MacroChem Corporation, FDA Drug Master File Number 10965.
65. H.P. Merkle, *Meth. and Find. Exp. Clin. Pharmacol.*, 1989, **11**, 135.
66. M.C. Musolf, in 'Transdermal Controlled Systemic Medications' Y.W. Chien (ed.), Marcel Dekker, New York, 1987, Chapter 4, p. 93.
67. T. Hille, in 'Pharmaceutical Skin Penetration Enhancement' K.A. Walters and J. Hadgraft (eds.), Marcel Dekker, New York, 1993, Chapter 15, p. 335.
68. K.L. Ulman, K.R. Larson, C.L. Lee and K. Tojo, *J. Contr. Rel.*, 1989, **10**, 261.
69. K. Tojo, C.C. Chiang, Y.W. Chien and Y.C. Huang, *Proceed. Intern. Symp. Control. Rel. Bioact. Mater.*, 1988, **15**, 412.
70. Adhesives Research, Inc., Transdermal Tech Brief 'Enhancer-Tolerant Adhesives'.
71. C.U. Ko, S.L. Wilking and J. Birdsall, *Pharm. Res.*, 1995, **12**, S-143.
72. H.J.C. te Hennepe, National Starch & Chemical B.V., Solvent Borne Pressure Sensitive Adhesives Development Group Info Bulletin, 'Solution Acrylic Polymers for Medical Applications', Issue 2, 1993.

Controlled Drug Release Using Hydrogels Based on Poly(ethylene glycols): Macrogels and Microgels

Neil B. Graham* and Jianwen Mao

UNIVERSITY OF STRATHCLYDE, GLASGOW G1 1XL, UK

1 INTRODUCTION

During the past twenty years, work on hydrogels based on crosslinked poly(ethylene glycols) in our own laboratories and elsewhere has led to the establishment of a number of controlled release products for pharmaceutical agents[1,2]. Some of these have been granted product licences in the UK and USA. Their use will almost certainly expand as essentially generic delivery systems for a variety of active agents. These materials are clearly able to provide a basis for many further and more advanced delivery technologies. Already, very advanced pulsed delivery systems are being developed[3-6] which will open up the entire sleeping period to timed delivery.

Hydrogels, in their simple monolithic form, do not easily lend themselves to the delivery of three groups of materials; those which are extremely water-soluble and so tend to release too quickly; those which are of very high molecular weight which tend to release too slowly for daily use; and those which require very high doses and for which it is difficult to achieve this high loading without surface-bolus dumping. These difficulties can be addressed by current developments which are the basis of this paper.

1.1 The Chemistry of Crosslinked Hydrogels Based on Poly(ethylene glycols)

Poly(ethylene oxide) polymers contain the following repeat unit; $-CH_2CH_2O-$, and are made commercially as two distinct groups of high and low molecular weight materials. All of these appear to have an exceptional degree of biocompatibility and have been utilised in a variety of pharmaceutical and biomedical applications. The low molecular weight materials are commercially available from a number of suppliers in grades of molecular weights up to $M_n = 8000$. Some higher molecular weights up to 20,000 are also available but are not commonly utilised. The low molecular weight materials which, usually, contain a hydroxyl on each end of their chains, are a well-known sub-group called the poly(ethylene glycol)s (PEG)s and are commonly used as excipients in pharmaceutical formulations. They have the general structure

---

* Correspondence Author

HO-$(CH_2CH_2O)_n$-H where "n" can be any number from 2 to approximately 200. The terminal hydroxyls can be reacted to form linear polymers of much higher molecular weights or alternatively used for their incorporation into crosslinked network structures. The W.R. Grace company market the group of derivatives known as Hypol^{TM} resins. These comprise poly(ethylene glycol)s of molecular weight around $M_n = 1000$ which have been reacted with an excess of diisocyanate and so endcapped with a free isocyanate group on each end of the molecule. The products are usually fluid and can be reacted with water to form a hydrophilic foam. The chemistry involved in the foaming is the reaction of the isocyanate with water as given in the example in Figure 1. Water-soluble or insoluble agents can be incorporated into the foamed structures and such combinations can be utilised for a relatively crude but effective sustained delivery device.

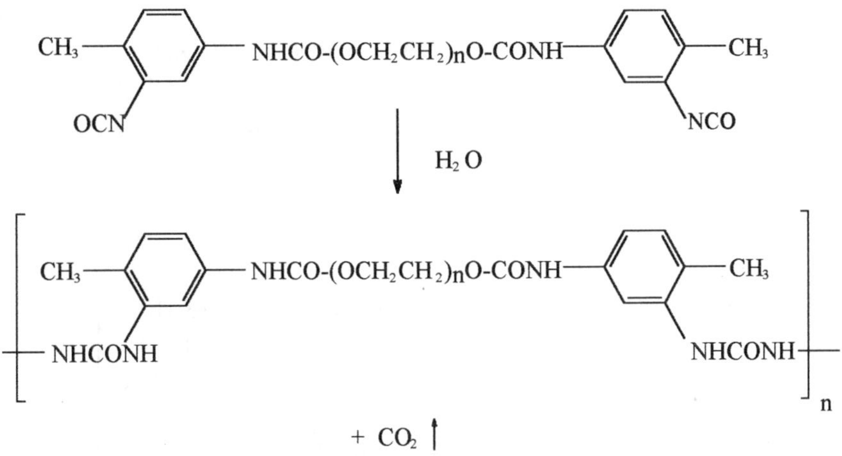

Figure 1. *Representation of the reaction of isocyanate end-capped poly(ethylene glycol)s with water to form a foamed structure*

Chemical crosslinks can be formed in such systems by either the deliberate incorporation of a trifunctional isocyanate end-capped polyol or by the incidental formation of biuret crosslinks. Biuret crosslinks are produced by the reaction of free isocyanate groups with already formed urea linkages.

In our work, a very different non-foamed monolithic hydrogel can be made in a very reproducible manner by the reaction of a diisocyanate with a selected poly(ethylene glycol) and a triol. Considerable skill and care is required in the preparation of these materials to the desired standards of pharmaceutical and regulatory reproducibility. It is easy to obtain a *polymer* of this class but not to obtain a ***precisely reproducible*** polymer. All composition factors must be tightly controlled along with the preparation conditions. By way of example; water contains two active hydrogens and has a molecular weight of 18. An impurity of 0.1 % w/w water in the formulation would contain the same number of active hydrogens as 33.3% w/w of a poly(ethylene oxide) of molecular weight $M_n = 6000$. Clearly it is not possible to control the stoichiometry nor to obtain reproducible polymer unless the hygroscopic poly(ethylene oxide) starting materials are thoroughly dried before their reaction to form polymer. The reaction of a diisocyanate, poly(ethylene glycol) and triol to form a crosslinked polyurethane hydrogel is illustrated below in Figure 2.

The polymers produced from the crosslinking of PEGs of M_n 2000 to 8000 differ from those of lower molecular weight in that they contain levels of crystallinity of typically around 50%. This, combined with the fact that PEG is a low glass transition temperature rubbery polymer, makes the material of a very tough consistency rather like polyethylene. This is in contrast to most dry xerogels which are glassy materials.

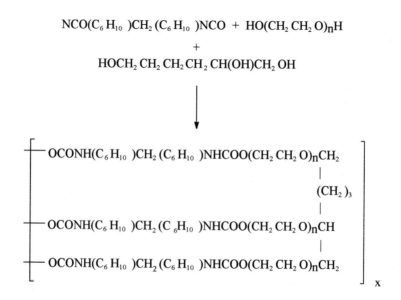

Figure 2. *Representation of a network arising from the reaction of poly(ethylene glycol) with biscyclohexylmethane-4,4'-diisocyanate and 1,2,6-hexanetriol*

Crosslinked polyurethane hydrogels of this type have also been found to swell with both water and organic solvents so that they can be readily charged with drug solutions[7]. Drying produces a material which in controlled geometries, provides constant drug release with many agents for the first fifty percent of the drug content. To obtain this desirable release profile, a monolithic slab of the material having minimum thickness of about 0.8mm must be used[8]. Such devices are very safe as the drug is dissolved in the carrier polymer and cannot be dumped. The first product of this type using prostaglandin E_2 for vaginal delivery to ripen the cervix prior to the onset of labour has received product licences in the UK and USA. A typical release profile for a monolithic lozenge shaped device of this type is shown in Figure 3.

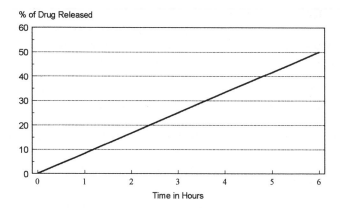

Figure 3. *A representation of the release pattern from a typical monolithic "lozenge" of 1 to 2 mm thickness containing 5 - 10 mg of prostaglandin E$_2$ releasing into buffer at pH7 at 37 °C.*

By the use of concentration profiling, the fraction of the contained active agent released at a constant rate can be increased to greater than 80% [9,10]. The simple monolithic system is however not a universal panacea, if any such exists. The diffusion coefficients of high molecular weight polymers and some proteins are so low in a non-microporous hydrogel that the release of such contained material takes an excessive time. This might be useful for a long acting implant but is not suitable for a daily dosage form. A microporous matrix is required for this latter and may be the most useful type of hydrogel for the release of proteins[11].

Very high loadings of very soluble drugs can be charged into hydrogels from appropriate solvents. However, this can result in a high concentration of drug at the hydrogel surface. A high surface concentration of drug usually provides an undesirable bolus dose and would probably lead to inadequate reproducibility of the commercial product. Hydrogels in small granular form containing very water soluble drugs tend to release their contents too fast to obtain release over several hours. A new ***microgel*** technology involving the discovery of a simple procedure to prepare soluble crosslinked polymers, and in particular those based on PEG, in the form of tiny (10-1000nm) particles, provides potentially attractive solutions to these problems.[12,13,14,15] The rest of this paper will be devoted to discussion of these materials and their applications.

1.2 Polyurethane Microgels as Novel Matrices

Polyurethane microgels were prepared by stoichiometrically polymerising different molecular weight poly(ethylene glycols) with diisocyanates and a trifunctional hydroxylic crosslinker in various organic solvents. According to gelation theory, such systems would normally form macrogels at complete reaction. However, work in our laboratories has shown that in the presence of thermodynamically good solvents, macrogelation can be completely circumvented. The resulting product is a "sol" of microgel which in most respects behaves like a solution. A number of important factors play an important role in the formation of the small crosslinked microgel particles. These factors include auto-steric stabilisation, the solubility parameter and hydrogen bonding ability of the solvent, the concentration of the monomers and the molar ratio of crosslinker.

Analytical techniques such as electron microscopy, NMR, GPC, FTIR and viscosity can be used to determine the structure of the microgel macromolecules. It is possible to synthesis microgels with a range of different properties and characteristics such as molecular weight, viscosity and water-solubility. This can be achieved by carefully selecting the appropriate monomers, molar ratio of crosslinker, solvent system and monomer concentration.

The microgels involved in this study are expected to have excellent biocompatibility and should therefore be suitable for use in the production of pharmaceutically acceptable controlled release formulations. Microgel powders can be easily dry-blended with drug substances and the well-blended mixtures can then be made into tablets by compression or by a melt-compression process developed in our laboratory. Tablet formulations can be prepared which are capable of either swelling or gradually dissolving in aqueous media over a period of time depending on the characteristics of the microgel used in the tablet fabrication. Sustained and often near zero-order drug release can be obtained from microgel tablets of this type.

2 EXPERIMENTAL

Microgels were prepared by the solution polymerisation of polyethylene glycol (M_n = 6000), 4,4'-dicyclohexylmethane diisocyanate(HMDI) and 1,2,6-hexanetriol(HT). The polymerisations were catalysed using anhydrous $FeCl_3$. Methyl ethyl ketone, acetone, diethyl ketone and dimethyl sulfoxide were used as solvents. The reactions were carried out in sealed serum bottles in an oven at 80°C. The completion of reaction was indicated by monitoring the disappearance of the characteristic isocyanate IR peak at $2215 cm^{-1}$ (normally ~24 hours). For each solvent and each composition, a maximum monomer concentration was found above which macrogelation occurred and below which microgels were obtained as the exclusive product. The microgels selected for use in this study were prepared in methyl ethyl ketone with monomer concentrations of 11.6g and 14.5g of monomers per 100 ml of solvent. The microgel compositions, molecular weights and water solubility are given in Table 1.

Table 1. *Characteristics and compositions of microgels used in this study.*

Sample	Monomer Concentration (g/100ml)	Hexanetriol Crosslinker (mole%)	PEG6000 (mole%)	Mw (GPC)	Solubility in Water
MH3-3	11.6	30.3	12.1	20,000	Yes
MH3-4	14.5	30.3	12.1	22,000	Yes
X2.5-3	11.6	30.3	12.1	160,000	No
X4-3	11.6	33.3	8.3	120,000	No

Caffeine(M.W.=194.19,), poly(N-vinyl pyrrolidone)(PVP) (M_w =10,000) and bovine serum albumin(BSA), fraction V powder from Sigma(M_w=66,000) were used as model drugs. The test formulations were prepared by blending the compound to be released with microgel granules which had been passed through a 422μ m sieve. The well blended mixtures were then compressed into tablets using a punching machine with a force of 5 tons. The tablet dimensions were ~13mm in diameter with a thickness of ~3mm. The approximate weight of each tablet was 0.48g. Tablets prepared using the microgel compositions X2.5-3 and X4-3 swelled in aqueous media without disintegration whereas tablets based on the microgel compositions MH3-3 and MH3-4 disintegrated after immersion in aqueous media.

The release of the model drug compounds from the various tablet formulations was investigated using a standard USP dissolution apparatus (Caleva Model 8ST). The release medium was 0.1N aqueous H_2SO_4 at a temperature of 37°C. The paddle speed was 60rpm. The release of drug was monitored by UV spectrophotometry (Cecil UV-Visible Spectrophotometer Model 5500). The analysis wavelengths used were 274nm, 215nm and 280nm for caffeine, PVP and BSA respectively.

3 RESULTS AND DISCUSSION

The size of the microgel particles was investigated by Transmission Electron Microscopy (TEM). A test specimen for TEM study was prepared by casting a drop of a very dilute microgel solution onto a carbon coated TEM sample holder and allowing the solvent to evaporate. The sample was then examined by TEM. Figure 4 shows a TEM photograph of the X2.5-3 microgel particles. The size of the particles was in the range of 60-100nm. If it is assumed that the particles are individual microgel molecules, then the molecular weight of each particle can be calculated from the particle volume and particle density. In order to do this, two assumptions must be made. Firstly, the microgel particles are assumed to be spherical and so the particle volume can be calculated using the measured particle diameters. Secondly, the density of the microgel particle is assumed to be equivalent to the density of the bulk poly(ethylene glycol) which comprises >70% of the microgel composition. The density of the bulk poly(ethylene glycol) is $1.13g/cm^3$. The molecular weight of the microgel particles using this method was found to be in the range of 0.55-2.50×10^9 Daltons. This molecular weight range for the microgel particles is clearly much higher than was obtained using GPC which indicated a molecular weight range of ~160,000 for the X2.5-3 microgels. It is possible that the actual density of the solvent-cast microgel particles is quite different to the density of the bulk poly(ethylene glycol). For this reason, the molecular weight values obtained using TEM should be treated with caution. It should also be emphasised that the particles observed by TEM have not been proven to be individual microgel molecules but may, in fact, be molecular aggregates. On the other hand, the GPC results are based on a calibration using linear poly(ethylene glycol) and poly(ethylene oxide) standards. The coil densities of the linear polymer standards in solution may be considerably lower than the densities of the crosslinked microgel particles. Consequently, the molecular weight values measured by GPC may be significantly too low and must also be treated with caution.

On completion of the reaction, the crosslinked poly(ethylene oxide) microgel particles are precipitated into petroleum ether (b.p. 60-80° C).

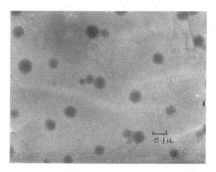

Figure 4. *Photograph of X2.5-3 microgel particles taken by Transmission electron Microscopy*

The precipitated particles are small crystalline aggregates of many individual microgel particles. Scanning electron microscopy (SEM) can be used to follow the growth of the crystalline microgel aggregates and the results indicate the formation of a dendritic open porous structure. When the dry microgel particles are compressed, the pores are closed up. However, when the particles are swollen in water, the pores are able to re-open to form water filled channels. If an active agent or drug is compressed in combination with the microgel particles, the release of the contained agent may follow the mechanism as illustrated in Figure 5.

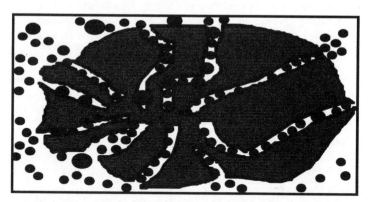

Figure 5. *Representation of the release of large molecules through channels in a compacted microgel (based on PEG) which has been subsequently swollen with water*

The fractional release of drug from a microgel tablet formulation is also influenced by the amount of drug loaded within the tablet.. For example, tablets were prepared from the water-soluble microgel composition MH3-3 and different amounts of the model drug compound, caffeine. It was found that the rate of release increased with increasing caffeine content. This may be explained by the fact that the dissolution and release of caffeine from the microgel tablets, results in the creation of further pores or micro-channels within the tablet structure. The extent of this increase in porosity is governed by the amount of caffeine loaded within the tablet. Higher caffeine loadings will, as a consequence of the increased porosity, facilitate higher rates of release. It was found that surprisingly high fractions of caffeine were released from the microgel/caffeine tablet formulations with essentially zero-order release rates. A sustained release of caffeine could be achieved for periods of between ~30 minutes and up to ~ 5 hours depending on the caffeine loading. The release profiles of the microgel/caffeine tablets are shown in Figure 6.

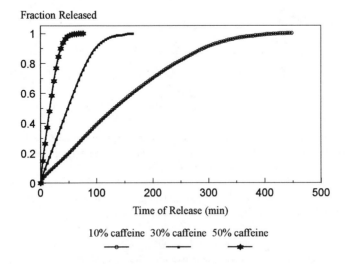

Figure 6. *Release of caffeine from a compacted water soluble hydrogel of MH3-3.*

A more prolonged release of caffeine was achieved by fabricating tablets using the microgel composition X2.5-3. The X2.5-3 microgel composition swells but does not dissolve when immersed in aqueous media. The release rate, as shown in Figure 7, was not as constant as the release rates obtained from the MH3-3 microgel tablet formulations. However, the release was now sustained for a period of approximately 9 hours. The X2.5-3 microgel tablet was swollen but still intact by the end of the caffeine release.

The release of the polymer PVP from the X2.5-3 microgel tablets was, as expected, slower than the release of caffeine. This is due to the higher molecular weight and lower diffusion coefficient of the PVP polymer chains compared with the relatively small caffeine molecules. The difference in the release rates of the caffeine and PVP from the microgel tablets was not as large as would have been expected if their release had been from a dense monolithic hydrogel matrix. This provides strong evidence that the release of active substances from microgel tablets is predominantly via water-filled pores rather than by diffusion through the aqueous phases within a water-swollen hydrogel continuum.

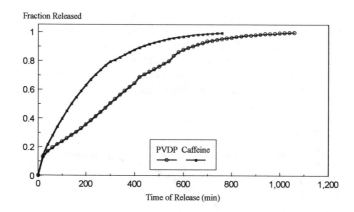

Figure 7. *Release of caffeine from a compacted water insoluble hydrogel of X2.5-3.*

The advancement of recombinant peptides and proteins as commercially available therapeutic agents has been so dramatic that some unique challenges have been raised for the development of appropriate drug delivery systems[16]. The controlled delivery of high molecular weight hydrophilic molecules such as proteins is a difficult objective. The use of hydrogel materials may provide a unique means for achieving this objective due to their hydrophilicity, relatively high permeability and biocompatibility.

The loading of proteins into a hydrogel matrix is difficult. Generally, two methods have been used to achieve such loading. The first method involves synthesising the hydrogel in the presence of the protein. The second method involves charging the protein into the hydrogel matrix by swelling the hydrogel in a solution of the protein. Both methods have serious drawbacks. The former could result in unknown side reactions between the proteins and the hydrogel components with the possible generation of allergenic or toxic products. It is also possible that the proteins may not be stable under the required reaction and processing conditions. The second method may not provide sufficiently high loadings of drug due to the low solubility of proteins in water or because the high molecular weight proteins may not easily diffuse into the hydrogel matrix. The drying process may lead to the formation of an undesirable surface layer of the active agent which may cause a high initial bolus dose. Charging proteins into hydrogels from organic solutions may also result in protein denaturisation.

Figure 8 shows the release of the protein BSA from an X2.5-3 microgel tablet containing 10%w/w of BSA compared to the release of PVP from an otherwise identical tablet. The release experiments were repeated three times and the results were shown to be reproducible. The release of protein was sustained over approximately 24 hours at a close to zero-order rate. The lower molecular weight PVP was released over a somewhat shorter period of approximately 17 hours.

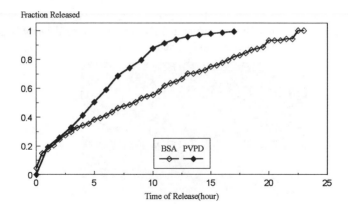

Figure 8. *Release of Bovine Serum Albumin from a compacted water insoluble hydrogel of X2.5-3*

The preparation of microgel tablets is a simple procedure and does not require the use of any added excipients. The advantage of using a polyurethane microgel matrix as a means of extending the controlled release of protein-based drugs such as insulin and growth factors is evident. However, the choice between the use of a dense crosslinked matrix or a microporous microgel matrix for the sustained delivery of low molecular weight peptides would need to be assessed on the merits of the individual cases and on the properties of the peptide to be delivered.

After release, the exhausted microgel tablets were examined by SEM. Water channels with sizes ranging from submicron to ~40μ m were observed in the tablets, Figure 9 and Figure 10.

Figure 9. *Scanning Electron Micrograph of the small holes n the exhausted compacted microgel matrix after swelling in water.*

The water channels are believed to provide a path through which the active agents are released. The dry microgel tablets are crystalline and essentially non-porous. When the microgel tablet is immersed in aqueous media, the initial swelling behaviour is similar to that of a non-porous monolithic hydrogel matrix. This provides an initially slow swelling system capable of providing a close to zero-order release profile in much the same manner as a monolithic hydrogel matrix. As the tablet continues to swell, micropores are opened up within the tablet and these pores are sufficiently large in size to allow the release of high molecular weight active agents.

Figure 10. *Scanning Electron Micrograph of the large holes in the exhausted compacted microgel matrix after swelling in water*

Normally, a drug is loaded into a dense crosslinked hydrogel matrix by swelling the matrix in a concentrated solution of the drug to be released. The loading processes are complicated and time-consuming. However, there are no such problems involved in the fabrication of microgel tablets.

Microgel matrices which are capable of forming micropores of significant size could also be used for the controlled delivery of microorganisms. Certain microorganisms can be obtained in dry powder form and could, in principle, be compressed together with microgels into compacts without damaging the microorganism. This type of sustained release system may provide interesting therapies in areas where natural microorganisms are currently utilised, e.g. intestinal and vaginal treatments

For use as long-term implants or as a multiplicity of granules for oral use, a modified hydrogel delivery system has been studied. This involves coating the compact of the microgel and the drug, which may be a tablet or granule configuration, with a water-impermeable polymer film. The polymer film is preferably a rubber which can stretch as the hydrogel swells with water. An example of this type of design is shown in Figure 11. In this example, the entry of water and the exit of the drug are both constrained by the size of a hole which is made in the coating. The release from these systems can be designed to take place over times from a day to many weeks and can, in principle, be tailored to any particular drug.

For implants, a related design has been evaluated for the sustained delivery of melatonin to sheep in microgram or milligram quantities per day. The design and illustration of the release device is shown in Figure 12. These devices originally used a dense crosslinked core which was not prepared from a microgel. The use of a microgel to replace the denser core should allow the prolonged release of therapeutically active proteins or perhaps microorganisms over long periods.

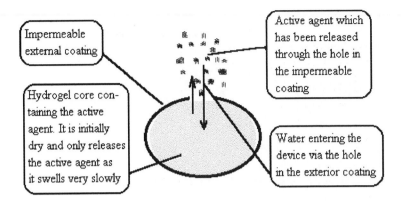

Figure 11. *Diagram of a hydrogel granule coated with a hydrophobic and impermeable polymer layer. A hole was provided for water to enter and for the contained active agent in the hydrogel to diffuse out. The granule can be of any size and the dimensions of the hole can be varied by design.*

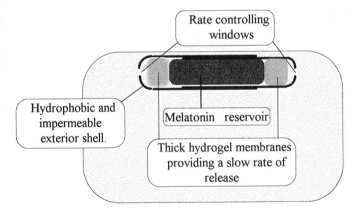

Figure 12. *Configuration of a release device used for the delivery of melatonin at rates of <1 mg/day for a period of two months.*

4 CONCLUSIONS

Polyurethane microgels have been used as a novel and promising matrix for controlled drug delivery. Water swellable microgels give more prolonged release than water soluble microgels. It has also been demonstrated that these microgels are very promising matrices for the controlled delivery of proteins and other active agents. The process for making such microgel tablets would be simple and need not involve either water or heating and so is a very practical procedure unlikely to cause damage to the active agent.

5 REFERENCES

1. N. B. Graham, Poly(ethylene glycol) gels and drug delivery. in "Poly(ethylene glycol) Chemistry: Biotechnical and Biomedical Applications", Ed. J. Milton Harris, Plenum Press, New York, 1992, 263.

2. J. Westgate and J. A. Williams, Evaluation of a controlled release vaginal prostaglandin E_2 pessary with a retrieval system for the induction of labour. *Journal of Obstetrics and Gynaecology*, 1994, **14**, 146.

3. A. Rashid, UK Patent.2230441B, Controlled Release Device, 1993.

4. M. E. McNeill, A. Rashid, and H. N. E. Stevens , UK Patent 2230442B, Controlled Release Capsules, 1994.

5. M. Bakhashaee, A. Rashid, and H. N. E. Stevens, UK Patent. 2241485B, Controlled Release Capsules, 1994.

6. J. Binns, M. Bakhashaee, H. N. E. Stevens and C. G. Wilson, Colon targeted release using the Pulsincap™ delivery system, *Proceed. Intern. Symp. Control. Rel. Bioact. Mater.*, 1994, Controlled Release Society Inc., **21**, 260.

7. N. E. Nwachuku, D. Walsh, and N. B. Graham, The interaction of poly(ethylene oxide) with solvents: 1. Preparation and swelling of crosslinked poly(ethylene oxide) hydrogel, *Polymer*, 1982, **23**, 1345.

8. M. E. McNeill and N. B. Graham, Properties controlling the diffusion and release of water-soluble solutes from poly(ethylene oxide) hydrogels 2. Dispersion in an initially dry slab, *J. Biomater. Sci, Polymer Edn.*, 1993, **5**, No. 1/2, 111.

9. M. E. McNeill and N. B. Graham, Properties controlling the diffusion and release of water-soluble solutes from poly(ethylene oxide) hydrogels 1. Polymer composition .*J. Biomater. Sci., Polymer Edn.*, 1993, **4**, No. 3, 305.

10. M. E. McNeill, and N. B. Graham, Morphine hydrogel suppositories: Device design, scale-up and evaluation in controlled release technology:, ACS Symposium Series 348, "Controlled-Release Technology, Pharmaceutical Applications", Eds. P. I. Lee and W. R. Good, Washington, 1987, Chapter 12, 158.

11. M. V. Badiger, and N. B. Graham, Porogens in the preparation of microporous hydrogels based on poly(ethylene oxides), *Biomaterials*, 1993, **14**, 1059.

12. N. B. Graham, and C. M. G. Hayes. Microgels 1 : Solution polymerization using vinyl monomers, *Macromolecular Symposium, Polymer Networks '94*, 1994, **93**, 293.

13. N. B. Graham, UK Patent 290264B, Polymerisation Process, 1984.

14. N. B. Graham, J. Mao, UK Patent Application 9405264.4, Microgels, 1994.

15. J. Mao, Internal Report, University of Strathclyde, 1993.

16. H Gehrke et al, Enhanced protein loading in hydrogels, *Proceed. Intern. Symp. Control. Rel. Bioact. Mater.*, 1993, **20**, 113.

Structural Investigations of the Monolayers and Vesicular Bilayers Formed by a Novel Class of Nonionic Surfactant

M. J. Lawrence[1], S. Chauhan[1], S. M. Lawrence[1], G. Ma[1], J. Penfold[2], J. R. P. Webster[2] and D. J. Barlow[1]

[1] DEPARTMENT OF PHARMACY, KING'S COLLEGE LONDON, MANRESA ROAD, LONDON SW3 6LX, UK

[2] RUTHERFORD-APPLETON LABORATORY, CHILTON, DIDCOT, OXFORDSHIRE OX11 0RA, UK

1 INTRODUCTION

Surfactants, when dispersed in aqueous solution, spontaneously aggregate to form a wide variety of equilibrium phase structures, many of which have potential for use as drug delivery vehicles.[1] Micelles, for example, have for many years been extensively studied as a means of increasing the apparent aqueous solubility of poorly water soluble drugs[2], while more recently, cubic phases have attracted considerable interest as a means of producing sustained release drug formulations[3]. Upon the addition of an oil to a binary surfactant/water mixture, the variety of equilibrium surfactant phase structures is frequently increased even further, with aggregates such as microemulsions and polymer-like reverse micelles[4] being formed. Many of these ternary phase structures, and in particular microemulsions, are themselves currently being intensively examined as potential drug carriers[5]. Unfortunately, despite this widespread interest in the use of surfactant systems as drug delivery vehicles, their commercial exploitation is not as widespread as might be expected. Indeed to the current authors' knowledge, there are only a handful of products on the UK market that can truly be classed in this way. These include Konakion and Vit A, that contain vitamins K and A respectively, solubilised in micelles of polyoxyethylated castor oil; Fungizone and Ambocil, which are colloidal and Sandimmun Neoral, in which cyclosporin is formulated as a "microemulsion pre-concentrate". All pharmaceutical products currently containing surfactants on the UK market leader used to solubilize poorly water soluble drugs either for acute conditions or for chronic, life-threatening diseases.

The disappointing lack of commercial exploitation of these potentially valuable surfactant systems as drug delivery vehicles arises firstly because of the paucity of pure, commercially available surfactants capable of forming the desired aggregates under the requisite conditions of temperature and concentration, and secondly, because the surfactants that are commercially available, and meet the stringent requirements for pharmaceutical quality control and form the required surfactant aggregate under the appropriate conditions, are frequently far too toxic, [6] and damaging to the environment. It is essential therefore that new, well-characterized non-toxic, preferably non-ionic biodegradable surfactants are designed and synthesized, if surfactant systems are going to realize their full potential as drug delivery vehicles. Indeed the serious problem encountered with the persistence of many of the so-called "non-toxic" nonionic surfactants in the environment and their resulting detrimental effects on the eco-chain means that many other industries, including the agrochemical and detergent, will also need to replace their existing range of surfactants with new, environmentally friendly alternatives.

2 DESIGN AND SYNTHESIS OF NEW PHARMACEUTICALLY ACCEPTABLE SURFACTANTS

2.1 Previous Work

To date, there has been very little effort expended in the design and synthesis of new surfactants specifically for the purposes of drug delivery. Much of the work that has been performed has been concerned with improving drug solubilization in micellar solutions. For example, during the eighties, Elworthy and co-workers attempted to improve the micellar solubilization of a wide range of drugs by designing new, non-ionic surfactants based on polyoxyethylene glycol that either produced very large micelles,[7,8] or encouraged solubilization of the drug in the hitherto relatively under-utilized hydrophobic core of the micelle.[8,9] (Note that, in the context of drug solubilization, unless the drug is extremely lipophilic (i.e. log P greater than 4), it is likely to be incorporated predominantly into the hydrophilic head group region of the micelle, in particular, the relatively dehydrated region close to the core.) Unfortunately both of the approaches of Elworthy and co-workers proved unsuccessful, although the results of their studies provide important clues as to how, in the future, to design new classes of surfactants with enhanced drug incorporation.

At about the same time as the studies performed by Elworthy and co-workers, other researchers[11] were investigating an alternative approach to increasing micellar solubilization, which involved replacement of the usual polyoxyethylene hydrophile present in most non-ionic surfactants with a heterocyclic amine-N-oxide group. Drug solubilization in the micelles produced by these amine-N-oxide surfactants showed a significant enhancement, sometimes by up to ten fold, over that achieved using the more conventional polyethylene oxide surfactants. Similar increases in solubilization have also been achieved for a range of steroidal drugs using the commercially available N,N-dimethyldodecylamine-N-oxide surfactants.[12] These studies suggest that it should be possible to design new surfactants with improved drug incorporation properties. However, since these studies, although a few new classes of surfactants have been synthesized, little, if any work has been specifically aimed at increasing the solubilization of drugs in micelles, or indeed any other surfactant aggregate.

More recently, Attwood, Booth and co-workers have initiated a programme to synthesize biodegradable polymeric non-ionic surfactants for the purposes of forming cubic phase structures useful as controlled drug release vehicles.[13] Although the group have not yet been successful in achieving their ultimate goal, they have succeeded in synthesizing a wide range of novel polymeric non-ionic surfactants that are worthy of further investigation [14,15] (Note, however that the work of this group is not specifically aimed at producing surfactants with enhanced drug solubilizing capacity.)

From the foregoing discussion it will be apparent, therefore, that there has been very little research concerned with the design and/or synthesis of new surfactants specifically for the purposes of drug delivery. This means that workers involved in pharmaceutical formulation are extremely limited in the number and range of surfactants that they are currently able to exploit.

2.2 Current Work

In the few cases where efforts have been made to design new surfactants, the approaches followed have been empirical or semi-empirical, rather than truly rational, with no a priori guarantee that the new surfactants synthesized would produce the required aggregate type. In an attempt to circumvent this serious limitation, we have embarked on a more rational program of surfactant design and synthesis, with the specific aim of producing molecules suitable for use in drug delivery. It is the primary goal of this work to be able to design well-characterized, non-toxic, biodegradable surfactants to produce the desired type of aggregate. In order to do this we have used molecular modelling techniques,[16] together with physico-chemical data, in an effort to predict which surfactants to synthesize in order to produce a particular aggregate.

3 COMPUTER AIDED DESIGN OF NEW SURFACTANTS

The advantages and the problems, associated with this strategy will be illustrated here with reference to our recent development of novel non-ionic dialkyl polyoxyethylene ether surfactants intended for the production of "stealth" (i.e. sterically stabilized) vesicles that exhibit prolonged *in vitro* and *in vivo* stability.[17] The production of stable vesicles for drug delivery is of paramount importance because, with a few exceptions,[18] vesicles do not generally form spontaneously and are therefore not thermodynamically stable, equilibrium structures.[19] As a consequence, they do not normally exhibit the long term stability required for their use as pharmaceutical drug delivery systems, and must in some way be stabilized. The requisite vesicle stability is normally achieved by covering the vesicle surface with long hydrophilic polymer chains.[17]

The novel surfactants used in the following study have the chemical structure illustrated in Figure 1. The novel surfactants were given the general formula $2C_nMPEG_{Mw}$, where n is the number of carbons in the alkyl chain and Mw is the molecular weight of the monomethoxy polyoxyethylene glycol head group. For the purposes of comparison the structure of zwitterionic phospholipid is also shown. As can be seen the structure of the novel surfactants is closely based on that of a phospholipid, with the distinction that both the hydrocarbon chains and the hydrophilic head groups are linked to the glycerol backbone via ether and not ester linkages. The reason for the use of the ether as opposed to ester linkage was two fold. The first was the ease of synthesis of ether containing lipid derivatives, as it is well known that transesterification of the acyl chains can occur during synthesis of ester lipids. In addition, ether lipids are more chemically stable than ester lipids. The second reason was the fact that data comparing ether and ester linked phospholipids suggested changing the linker moiety has very little effect on the physico-chemical properties of the vesicles formed. The replacement of the phosphorylcholine head group should provide the novel surfactants with the potential benefits afforded by a steric barrier, namely an increase in both the *in vitro* and *in vivo* stability of the vesicular structures.[17] Providing the steric barrier as an integral part of the surfactant constituting the vesicle should overcome many of the problems experienced with other methods of providing sterically stabilized vesicles, e.g. physically coating pre-formed vesicles with polymers,[22] chemically coupling polymer to pre-formed vesicles[23] and the addition of micelle forming lipid derivatives[24] or polymeric surfactants.[25]

Finally, we note that this class of novel vesicle forming surfactants should, because of their structural similarity to zwitterionic surfactants and very low critical micelle concentration, exhibit a relatively low toxicity. Indeed, preliminary experiments have demonstrated that these surfactants have little or no toxicity in the macrophage-like J774 cells.[26]

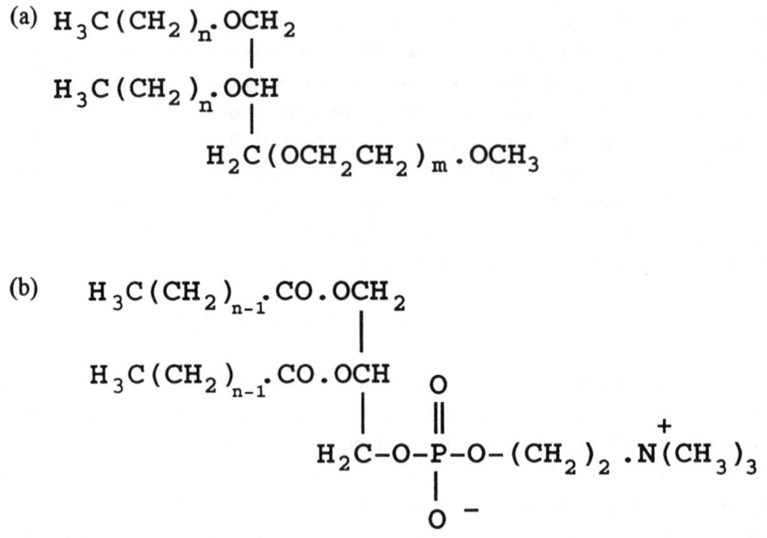

(a) $H_3C(CH_2)_n.OCH_2$

$H_3C(CH_2)_n.OCH$

$H_2C(OCH_2CH_2)_m.OCH_3$

(b) $H_3C(CH_2)_{n-1}.CO.OCH_2$

$H_3C(CH_2)_{n-1}.CO.OCH$

$$H_2C-O-\overset{O}{\underset{O^-}{\overset{\|}{P}}}-O-(CH_2)_2.\overset{+}{N}(CH_3)_3$$

Figure 1 *General structural formulae for (a) novel non-ionic surfactants and (b) phosphatidylcholines.*

To a first approximation it should, by analogy with the relationship between chemical structure and aggregate formation, be possible to predict which of the novel non-ionic surfactant chemical structures would be most likely to produce vesicles. For example, surfactants with a short hydrophilic head group and long hydrophobic chains would be expected to produce vesicles, whereas those with longer hydrophilic head groups and/or short hydrophobic chains may be expected to form micelles or hexagonal phases. However, when it is considered that the synthetic route required to produce the novel surfactants having symmetric hydrocarbon chains involves four stages[27] while the pathway needed to produce the asymmetric surfactants comprises six stages,[27] it is clear that it would be highly desirable (even if only from the point of view of economy) to know to a higher degree of probability which surfactants to synthesize in order to produce vesicles.

3.1 Use of VESICA

With this aim, we have developed the VESICA program, which provides for VEsicle Simulation and Computer Analysis.[16] As input, the program requires estimates of the hydrophobe chain length and volume of the surfactant, both of which can be readily calculated, together with either the surfactant head group area or the number of surfactant molecules contained in the vesicle bilayer.[16] For the type of surfactant molecule in question, by far the easiest piece of information to obtain is the area per molecule. This is a value routinely obtained from the force area isotherms, using a Langmuir trough (determined as the intercept formed between the abscissa and the tangent to the curve taken at the collapse point).

3.1.1 Langmuir Trough Data. Figure 2 shows the force area isotherm obtained for a spread monolayer of the surfactant, $2C_{18}MPEG_{550}$ using a Nima Trough at 25°C. The isotherm obtained differs considerably from that seen with phospholipid monolayers in that it exhibits a very expanded nature at very high areas per molecule, suggesting that the polyoxyethylene chains, although relatively short, are occupying a large area. At intermediate to low areas per molecule the isotherm exhibits an "apparent" region of condensation of the monolayers, i.e. a transformation from an expanded to a condensed film, before its eventual collapse at a pressure of approximately 56 mNm^{-1}.

All of the novel non-ionic surfactants examined gave rise to force area isotherms showing the same features, although the actual areas per molecule at which the various features occurred varied very slightly. At surface pressures below the "apparent" region of condensation, the monolayers could readily be decompressed, without exhibiting any hysteresis. When the monolayers compressed in the "condensation" regions were decompressed, hysteresis was observed, although the film eventually recovered to its initial values of surface pressure and area per molecule. The fact that the monolayer film could be decompressed and recompressed to yield the same isotherm suggested that very little, if any, of the surfactant was being lost to the aqueous sub-phase upon compression.

Interestingly, spread monolayers of distearylphosphatidylethanolamine containing less than 9 mol% of the surfactant polymer DSPE-EO$_{45}$ [28] also exhibit relatively high surface pressures at large areas per molecules, suggesting that the hydrophilic polymeric head group in these films also occupy large areas, as a consequence of the repulsive lateral steric interaction of the polyoxyethylene chains. In these mixed systems however no region of condensation is observed, probably due to the fact that only 9 mol% of the molecules contain polyoxyethylene glycol as their head group. Interestingly, when the mixed DSPE-DSPE-EO$_{45}$ films were recompressed after decompression from surface pressures below their collapse point, few molecules appeared to be lost from the monolayer (this result was surprising as DSPE-EO$_{45}$ is known to form micelles and may be expected to be lost to the aqueous sub-phase upon compression). Similar results have been seen by Burner et al.[29]

The values of the areas per molecule that were obtained from the Langmuir trough data for the novel non-ionic surfactants are given in Table 1. For the purposes of comparison the corresponding values for distearylphosphatidylcholine (DSPC) and distearylphosphatidylethanolamine (DSPE) are also listed. (The areas per molecule obtained for other phospholipids in the two series will not be expected to differ significantly from those obtained for DSPC and DSPE and are therefore not shown.)

It can be seen from Table 1 that the areas per molecule calculated on the basis of the surface tension data were, in all cases, less than that obtained for DSPC, although generally slightly larger than that observed for DSPE. When the values given in Table 1 were employed as input for VESICA modelling of the surfactant aggregates, the program predicted that the novel non-ionic surfactants would not form vesicles, but would instead favour the formation of some inverse structure. This result was obtained as a consequence of the fact that the measured head group area was too small to allow the formation of the requisite bilayer structures needed for vesicle formation. Indeed it is well known that DSPE and other phosphatidylethanolamines do not form vesicles but rather reverse micelles!

This prediction was totally unexpected since (as detailed below) a considerable amount of physico-chemical data had previously been collected that strongly supported the formation of vesicles by the novel non-ionic surfactants.[27]

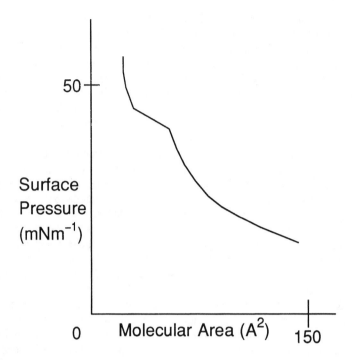

Figure 2 *Force area isotherm at 25°C for $2C_{18}MPEG_{550}$*

Table 1 Areas per molecule of the novel surfactant determined at the collapse pressure and 25°C

Surfactant Structure	Area per molecule ($Å^2$)
$2C_{22}MPEG_{750}$	52
$2C_{18}MPEG_{750}$	45
$2C_{16}MPEG_{750}$	55
$2C_{18}MPEG_{550}$	55
DSPC	66[30]
DSPE	43[28]

    3.1.2 Physico-Chemical Evidence for Vesicle Formation. Electron microscopy (using both freeze fracture and staining techniques) had given evidence for the formation of closed vesicular structures, and further support for these aggregates was provided by the results of other investigations, such as the encapsulation of the water soluble drug, sodium stibogluconate, together with sizing studies using photon correlation spectroscopy.[27]

    Interestingly, however, scanning tunnelling microscopy, while undoubtedly demonstrating the existence of spherical aggregates, suggested that the minimal sized vesicles formed by the novel non-ionic surfactants were considerably smaller than those formed by zwitterionic phospholipids such as DSPC. Values of about 12 nm were determined for vesicles produced by $2C_{18}MPEG_{550}$ as opposed to diameters of about 22 nm estimated by light scattering measurements for phospholipid vesicles.[32] Indeed, we had previously observed that the vesicles formed by the non-ionic surfactants using a wide variety of techniques (including the stable plurilamellar method, the thin film method followed by sonication and reverse phase evaporation) had routinely yielded vesicles smaller than those formed by the zwitterionic phospholipids with comparable hydrocarbon chain lengths. For example, vesicles formed by the novel surfactants using bath sonication were of the order of 200 nm diameter, while vesicles of diameters of greater than 1000 nm were obtained from a range of phosphatidylcholines prepared under identical conditions.[31]

As stated earlier, these observations confirming the presence of vesicles were at variance with the results of the VESICA modelling of the non-ionic vesicles using the areas per molecule obtained from the Langmuir trough experiments. In fact the calculations showed that the production of minimal sized non-ionic vesicles of smaller size than the phospholipid vesicles would require a much larger head group area than that seen for the phospholipid molecules, and yet the areas per molecule determined for the non-ionic surfactants were significantly smaller.

In order to try and explain the discrepancy between the results of the modelling using the data obtained from the Langmuir trough experiments and the observation that vesicles were indeed formed, we performed neutron specular reflectance measurements on the surfactant $2C_{18}MPEG_{550}$ adsorbed at an air-water interface.[33,34,35] In addition to providing an area per molecule for the non-ionic surfactants when adsorbed at an air-water interface, the neutron reflectance measurements yield the most detailed information currently available as to the structure of the adsorbed surfactant monolayer.[36]

3.1.3. Neutron Specular Reflectivity Measurements. The experiments were conducted using the CRISP spectrometer at the Rutherford-Appleton Laboratories, Didcot, Oxon. The reflectivity data were obtained as a function of surface pressure over the range 20-40 mNm^{-1}, deliberately working therefore at surface pressures below the regime over which the "apparent" condensation of the monolayer film had been observed. The data were obtained using the alkyl chain deuterated version of $2C_{18}MPEG_{550}$ on D_2O, and on air-contrast-matched water (ACMW). The results of the data analysis using the optical matrix method of Born and Wolf [37] assuming a single uniform layer fit are shown in Table 2. The analyses indicate that at the relatively low surface pressures of 11, 20, 28 and 34 mNm^{-1}, the adsorbed surfactant monolayer can be adequately modeled as a single uniform layer,[35] with the calculated low scattering length density of the monolayer indicating that there is a significant degree of mixing of the deuterated alkyl chains and protonated head groups. Additional data were subsequently obtained using the fully protonated version of the surfactant on D_2O, at the surface pressures of 20 and 28 mNm^{-1}, and it was then possible to combine these data with those obtained using the chain deuterated surfactant to permit more detailed optical matrix analysis assuming a two-layer fit. The combined data proved to be quite well modelled using the two-layer fit, with the analysis suggesting that one of the sub-layers within the monolayer consisted of approximately 60% of the oxyethylene groups immersed in the aqueous sub-phase, with the other made up of the remaining oxyethylene groups mixed with the glycerohydrocarbon moieties. This head group/hydrocarbon chain mixing has not been observed with monolayers formed by zwitterionic phospholipids, possibly because of the highly charged nature of their head group. Interestingly, however, it is not the first time that mixing of the oxyethylene groups of a surfactant head group with its hydrocarbon chains has been postulated. For example, Lu et al.,[36] have also observed a mixing of the hydrocarbon chains and polyoxyethylene head groups in monolayers formed by single chain polyoxyethylene ether surfactants. Furthermore Arnason and Elworthy[7] and Robson and Dennis[38] have also postulated, albeit for different reasons, that this mixing can occur in micelles formed by polyoxyethylene surfactants. Indeed while it is well known that polyoxyethylene glycols do not mix with hydrocarbons, their methylated equivalents are known to be completely miscible. It is not yet clear which part(s) of the ethylene oxide head groups of $2C_{18}MPEG_{550}$ molecules mix with the hydrocarbon chain region. As there is no free terminal hydroxyl present in the novel non-ionic surfactants tested, it is feasible that the terminal portion of the surfactant can curl round and penetrate the hydrophobic region. Alternatively the oxyethylene groups close to the hydrocarbon chains may protrude into the hydrophobic region. However, due to the short ethylene oxide chain lengths present there may be good entropic reasons why one of the options is more likely than the other.

Table 2 *Surface pressure variation in $2C_{18}MPEG_{550}$ monolayer structure*

Surface Pressure mNm^{-1}	Area per molecule (Å^2)	Number of water molecules per oxyethylene group
11	186 ($\pm$7.7)	3.1
20	153 ($\pm$2.7)	1.5
28	121 ($\pm$2.3)	1.9
34	99 ($\pm$2.7)	2.9

More detailed structural information on the monolayer formed by $2C_{18}MPEG_{550}$, at surface pressures less than 40 mNm^{-1}, has recently been obtained using a range of partially deuterated forms of this surfactant, applying the kinematic approximation[36] in the analyses of the resultant reflectivity profiles. The same observation is recorded that a significant proportion of the polyoxyethylene chains are mixed with the hydrophobic region of the surfactant.[39]

Upon compression of the monolayer to a surface pressure of 40 mNm^{-1}, the errors incurred in the modelling of the data using the two-layer fit become much pronounced, indicating that there is a far greater degree of multilayering at the interface.[33] Furthermore, there is also the development of a pronounced off-specular reflectance, suggesting the formation of islands of multilayers. Upon decompression these islands appear to remain for some time and are the source of the hysteresis observed in the adsorption isotherm.

It is clear from the results of the analysis of the neutron scattering that $2C_{18}MPEG_{550}$ behaves as a monolayer only at surface pressures below 40 mNm^{-1}. At greater surface pressures patches of multilayers, probably bilayers appear to be formed. This observed multilayering accounts for the "apparent" condensation observed in the pressure area isotherm, and also explains the very low areas per molecule obtained at the collapse point. The formation of these multilayers is undoubtedly a consequence of the fact that the polyoxyethylene head groups and the hydrocarbon chains can become readily mixed. It is therefore obvious that the wrong values of area per molecule were used in the original calculations of vesicle structure. Recalculation of the vesicle model using a more appropriate value of area per molecule (120 Å^2) obtained from extrapolation of the neutron reflectivity data,[35] leads to the prediction that the minimum sized vesicles will have an overall diameter of approximately 16 nm, satisfyingly close to that measured by scanning tunneling microscopy (i.e. 12 nm) and much smaller than the dimensions recorded for the corresponding phosphatidylcholine vesicles. Figure 3 compares representative conformations of the non-ionic surfactant in $2C_{18}MPEG_{550}$ the inner and outer monolayers of a minimal sized vesicle with that of DSPC.

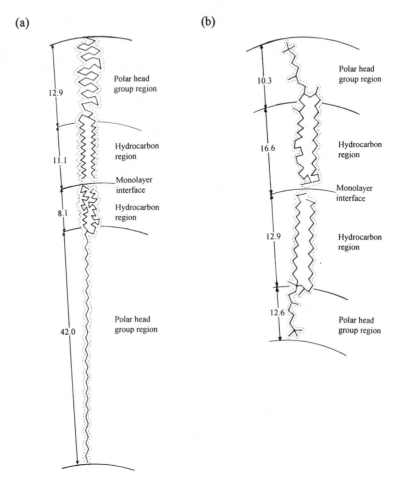

Figure 3 *Representative conformations of (a) the non-ionic surfactant $2C_{18}MPEG_{550}$ and (b) DSPC in the inner and outer monolayers of a minimal sized vesicle. Dimensions in Å.*

Although the neutron reflectivity experiments have so far been performed on only one of the novel surfactants in this series, it is not unreasonable to assume (given the similarities in their surface pressure isotherms), that the related surfactants would also yield large areas per molecule if examined by neutron reflectivity; and indeed since the other surfactants studied all possess longer hydrophilic head groups, even larger areas per molecules would be expected.

3.1.4. Molecular Modelling. Modelling of the vesicle produced by $2C_{18}MPEG_{550}$ using the data obtained from neutron scattering studies has shown that its structure is quite different from that of a zwitterionic phospholipid vesicle.[35] In particular the vesicles are about 2/3 the size of those produced by the phospholipid counterpart (diameter 158 vs. 276 Å), although because of their large polyoxyethylene head group the total bilayer thickness is about 3/2 greater than that seen for DSPC. The small vesicle size and large head group leads to a pronounced difference in the volume of the trapped aqueous cavities of the two vesicles, with vesicles of $2C_{18}MPEG_{550}$ entrapping only 0.43 nm^3 of water, while vesicles produced by DPSC have volumes around 2600 nm^3. Another important difference is the relative levels of head group hydration. The approximate head group hydration observed for $2C_{18}MPEG_{550}$ was about 2H$_2$O per oxyethylene unit. These hydration data are the first ever reported for a double chain surfactant bearing a polyoxyethylene head group, and are in good agreement with those reported for single chained polyoxyethylene surfactants, i.e. 1-3 H$_2$O.[40] Assuming that the same level of hydration applies for curved $2C_{18}MPEG_{550}$ monolayers, then in the unilamellar vesicles formed by this surfactant the outer polyoxyethylene head group will occupy a volume of 10^6 Å^3, accounting for just under half of the volume of the entire vesicle.

The observations made from the modelling study aided in the understanding of various of the physico-chemical properties of the vesicles. For example, when the encapsulation efficiency of the non-ionic vesicles towards the water soluble drug, sodium stibogluconate was examined the results were disappointing in comparison to the encapsulation values quoted for phospholipid vesicles.[27] From the modelling study, this can be rationalized in terms firstly of the smaller size and consequently, lower entrapped water volume of the non-ionic vesicles, and secondly the small amount of free water available to entrap the water soluble drug which would be reduced by the higher level of hydration of the polyoxyethylene chains, viz. at least 24 per $2C_{18}MPEG_{550}$ molecule as compared to approximately 12 for each phosphorylcholine head group.[41]

Table 3 *Stability of vesicle preparations with respect to size*

Surfactant	Immediately	2 weeks	1 month	2 month
$2C_{16}MPEG_{550}$	143 nm	150 nm	148 nm	144 nm
$2C_{16}MPEG_{750}$	131 nm	127 nm	136 nm	140 nm
$2C_{18}MPEG_{550}$	112 nm	117 nm	121 nm	118 nm
$2C_{18}MPEG_{750}$	115 nm	98 nm	103 nm	103 nm

The vesicles have been shown to be very stable with respect to time over periods of several months (Table 3). To date most workers have discussed the effect of the presence of polyoxyethylene glycol on vesicle stability in terms of the length of the polyoxyethylene chains, tending to ignore the fact that the *density* of the head group coating is also important.[17] In the vesicles produced by the non-ionic surfactants, the polyoxyethylene chains are considerably shorter than has previously been shown to stabilize vesicles, i.e. a molecular weight of about 550 as opposed to 1900-5000.[17,42] However in these cases the stabilizing amphiphiles represent only a relatively small fraction (5-10 mol %) of the total number making up the vesicle, and so cover only a correspondingly small fraction of the vesicle surface. In contrast, in the vesicles prepared in this study, the whole surface is covered with heavily hydrated short polyoxyethylene chains. Calculations for the minimum size vesicles show that the outer hydrated polyoxyethylene layer will provide a very highly substantial steric barrier which is likely to confer a high degree of *in vitro* and *in vivo* stability, with the conclusion reached that these class of surfactants offer significant advantages for vesicular drug delivery.

4 REFERENCES

1. M. J. Lawrence, *Chem. Soc. Rev.*, 1994, **23**, 417.
2. A. T. Florence, In "Techniques of Solubilization of Drugs", editor S. H. Yalkowsky, Marcel Dekker, Inc., New York, 1981.
3. S. Engstrom, L. Lindahl, R. Wallin and J. Engblom, *Int. J. Pharm.*, 1992, **86**, 137.
4. R. Scartazzini and P. L Luisi, *J. Phys. Chem.*, 1988, **92**, 829.
5. M. J. Lawrence, *European J. Drug Meta. and Pharmacokin.*, 1994, **19**, 257.
6. D. Attwood and A. T. Florence, "Surfactant Systems: Their Chemistry, Pharmacy and Biology", Chapman and Hall, London, 1983.
7. T. Arnarson and P. H. Elworthy, *J. Pharm. Pharmacol.*, 1980, **32**, 381.
8. P. H. Elworthy and M. S. Patel, *J. Pharm. Pharmacol.*, 1982, **34**, 543
9. D. Attwood, P. H. Elworthy and M. J. Lawrence, *J. Chem. Soc. Faraday Trans. I*, 1986, **82**, 1903-1910.
10. D. Attwood, P. H. Elworthy and M. J. Lawrence, *J. Pharm. Pharmacol.*, 1989, **41**, 585-589.
11. M. J. Lawrence and F. Devinsky, *J. Pharm. Pharmacol.*, 1988, **40**(suppl), 125P
12. C. Satra, M. Thomas and M. J. Lawrence, *J. Pharm. Pharmacol.*, 1995, In Press.
13. L. Martini, D. Attwood, J. H. Collett, C. V. Nicholas, S. Tanodekaew, N. J. Deng, F. Heatley and C. Booth, *J. Chem. Soc., Faraday Trans.*, 1994, **90**, 1961.
14. A. D. Bedells, R. M. Arafeh, Z. Yang, D. Attwood, J. C. Paget, C. Price and C. Booth, *J. Chem. Soc., Faraday Trans.*, 1993, **89**, 1243.
15. Z. Yang, S. Pickard, N. J. Deng, R. J. Barlow, D. Attwood and C. Booth, *Marcomolecules*, 1994, **27**, 2371.
16. S. M. Lawrence, M. J. Lawrence and D. J. Barlow, *J. Mol. Graphics*, 1991, **9**, 218.
17. M. C. Woodle and D. D. Lasic, *Biochim Biophys. Acta*, 1992, **1113**, 171.
18. M. Corti, L. Cantu and P. Salina, *Adv. Colloid Interface Sci.*, 1991, **36**, 153.
19. V. Kumaran, *J. Chem. Phys.*, 1993, **99**, 5490.
20. R. Evstigneeva, In "Phospholipids Handbook", editor G. Cevc, Marcel Dekker Inc., New York, 1993.
21. H. Eibl, *Agnew. Chem. Int. Ed.Engl.*, 1984, **23**, 257.
22. M. Jamshaid, S. J. Farr, P. Kearney and I. W. Kellaway, *Int. J. Pharm.*, 1988, 48, 125.
23. J. Senior, C. Delgado, D. Fisher, C. Tilcock and G. Gregoriadis, *Biochim. Biophys. Acta*, 1991, **1062**, 77.
24. M. C. Woodle, K. K. Mathay, M. S. Newman, J. E. Hidaya, T. R. Collins, C. Redeman, F. J. Martin and D. Papahadjopoulos, *Biochim. Biophys. Acta*, 1992, **1105**, 193.
25. B. Kronberg, A. Dahlman, J. Carlfors, J. Karsson and P. Artuson, *J. Pharm. Sci.*, 1990, **79**, 667.
26. S. Chauhan, PhD Thesis, University of London, 1992.
27. M. J. Lawrence, S. Chauhan, S. M. Lawrence and D. J. Barlow, *S.T.P. Pharma Sciences*, 1996, In Press.

28.   H. Burner, M. Winterhalter and R. Benz, *J. Colloid Interface Sci.*, 1994, **168**, 183.
29.   T. L. Kuhl, D. E. Leckband, D. D. Lasic and J. N. Israelachvili, *Biophys. J.*, 1994, **66**, 1479.
30.   B. A. Lewis and D. M. Engleman, *J. Mol. Biol.*, 1983, **166**, 221.
31.   C. J. Roberts, M. C. Davies, S. J. B. Tendler, S. M. Lawrence, M. J. Lawrence, S. Chauhan and D. J. Barlow, *J. Pharm. Pharmacol.*, 1991, **43**(suppl), 98P.
32.   B. A. Cornell, G. C. Flecther, J. Middlehurst and F. Separoric, *Biochim. Biophys. Acta*, 1981, **642**, 375.
33.   G. Ma., D. J. Barlow, J. Penfold, J. Webster and M. J. Lawrence, *J. Pharm. Pharmacol.*, 1994, **46**, 1085.
34.   G. Ma., D. J. Barlow, J. Penfold, J. Webster and M. J. Lawrence, *J. Pharm. Pharmacol.*, 1995, **47**, In Press.
35.   D. J. Barlow, G. Ma, M. J. Lawrence, J. Penfold and J. R. P. Webster, *Langmuir*, 1995, **11**, In Press.
36.   J. R. Lu, Z. X. Li, R. K. Thomas, E. J. Staples, I. Tucker and J. Penfold, *J. Phys. Chem.*, 1993, **97**, 8012.
37.   M. Born and E. Wolf, "Principles of Optics", 5th Edition, Pergamon Press, Oxford, 1973.
38.   R. J. Robson and E.D. Dennis, *J. Phys. Chem.*, 1977, **81**, 1075.
39.   D. J. Barlow, G. Ma, J. Penfold, J. R. P. Webster and M. J. Lawrence, In Preparation.
40.   C. Tanford, Y. Nozaki and M. F. J. Rhodes, *J. Phys. Chem.*, 1997, **81**, 1555.
41.   D. M. Small, *J. Lipid Res.*, 1967, **8**, 551.
42.   D. Napper, In "Colloidal Dispersions", editor J. W. Goodwin, Royal Society of Chemistry special publication 43, London, 1982.

Non-ionic Surfactant Vesicles and Colloidal Targeting Delivery Systems: the Role of Surfactant Conformation

Graham Buckton

CENTRE FOR MATERIALS SCIENCE, SCHOOL OF PHARMACY, UNIVERSITY OF LONDON, 29–39 BRUNSWICK SQUARE, LONDON WC1N 1AX, UK

1 INTRODUCTION

The use of non-ionic surfactants is now widespread in the pharmaceutical sciences. Current applications include the stabilisation of suspensions and creams, the preparation of vesicles and the coating of surfaces to facilitate drug targeting. In this review two applications will be considered for which it will be shown that it may be advantageous to consider surfactant conformation when using such materials. The two areas that are considered are the targeting of coated colloidal particles and the toxicity of surfactant vesicles.

2 THE TARGETING OF COLLOIDAL PARTICLES

2.1 How are particles cleared from the blood stream?

The adsorption of plasma proteins to exogenous particles is an important part of the defence mechanism of the body. Foreign materials are cleared from the body by the mononuclear phagocytic system (MPS). Proteins in the serum which promote phagocytosis are called *opsonins* (the word opsonin comes from the Greek, and means a relish, a seasoning or a sauce; it has often been quoted as being "the preparation of a meal"). Opsonisation is a process whereby materials adsorb to a foreign surface in such a way as to apparently prepare the surface for recognition as foreign, and thus for phagocytosis by the MPS. Much of the workings of the MPS still need to be understood, but it is probable that it is the conformational changes of opsonins that result in the plasma clearance.

2.2 The use of poloxamer surfactants to change clearance from the blood stream

Illum and Davis [1] showed that the adsorption of poloxamers onto the surface of polystyrene latex microspheres would produce a hydrophilic surface with low surface charge. The formation of such a surface limits the adsorption of opsonins and thus allows extended circulation in the plasma, as well as giving the possibility of distribution to different organ sites. Poloxamers are a-b-a block copolymers of the poly(oxyethylene)-poly(oxypropylene)-poly(oxyethylene) type. The structural details of the poloxamer surfactants are given in Table 1. It was found that if the polystyrene was uncoated or

Table 1.   Structural details of a.b.a block copolymers of poly(oxyethylene) (a)
(POE), and poly(oxypropylene) (b) (POP).

Poloxamer	Pluronic*	Molecular weight	POP units	POE units
101	L31	1100	16	2 x 2
105	L35	1900	16	2 x 11
108	F38	5000	16	2 x 46
122	L42	1630	21	2 x 5
123	L43	1850	21	2 x 7
124	L44	2200	21	2 x 11
181	L61	2000	30	2 x 3
182	L62	2500	30	2 x 8
184	L64	2900	30	2 x 13
185	P65	3400	30	2 x 19
188	F68	8350	30	2 x 75
215	P75	4150	35	2 x 24
217	F77	6600	35	2 x 52
235	F85	4650	39	2 x 27
237	F87	7700	39	2 x 62
284	P94	4600	47	2 x 21
331	L101	3800	54	2 x 7
333	P103	4950	54	2 x 20
334	P104	5850	54	2 x 31
335	P105	6500	54	2 x 38
338	F108	14000	54	2 x 128
407	F127	11500	67	2 x 98

(* Pluronic is a trade mark of ICI)

coated with poloxamer 188 the particles were mostly deposited in the liver and spleen. The only significant difference between uncoated and 188 coated particles was that the amount found in the lung increased for the surfactant coated material. If, however, the particles were coated with poloxamer 338 a very different distribution was seen (rabbits were used as the animal model). The 338 coated particles showed a lower tendency to concentrate in the liver and spleen, and more of the particles were seen to be circulating in the blood and deposited in the heart, lung and carcass. Illum and Davis[1] described the adsorption of the block copolymer to the latex, which will be by attachment of the poly(oxypropylene), with the poly(oxyethylene) protruding from the surface (shown schematically in Figure 1). Illum and Davis[1] stated that the mechanism for the difference in organ distribution between 188 and 338 coated particles was either due to different binding affinities for the two poloxamers, or due to steric stabilisation. The basis of the steric stabilisation argument was that the hydrophilic poly(oxyethylene) chains, would protrude to different lengths from the surface for different adsorbed poloxamers, and thus have a range of effects. The steric stabilisation concepts are based on the summation of attractive and repulsive energies between two materials as a function of distance, as expressed by the DLVO theory. Illum *et al.*[2] presented data on more poloxamers, in relation to their uptake by mouse peritoneal macrophages. In this publication, reference

Figure 1. Schematic representation of the adsorption of a poloxamer surfactant onto a hydrophobic surface.

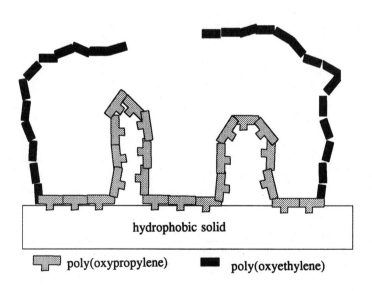

The poly(oxypropylene) units (shown as white blocks) bind to the hydrophobic surface with trains of contact separated by loops. The poly(oxyethylene) (shown as shaded blocks) protrudes from the surface.

was made to the fact that different concentrations of surfactant resulted in different coating thicknesses on the polystyrene latex. The effect of surfactant concentration is something which has not been addressed thoroughly, as it has been assumed by many workers that any concentration above that which was reported to give a plateau in the adsorption isotherm (by Kayes and Rawlins [3]) will produce equivalent surfaces: this need not be the case. A plot of the phagocytic uptake observed by Illum *et al.* [2] as a function of measured coating thickness is presented in Figure 2. Given the relationship in Figure 2, which shows that as the coating layer thickness increases, so the phagocytic uptake decreases, it is easy to see why the hypothesis that the effect was due to steric stabilisation flourished. However, if the molecular weight of the hydrophobe of the block copolymer is taken as a very crude assessment of the possible binding strength to the latex, it is possible to produce Figure 3 from the data of Illum *et al.* [2]. The linear correlation to the fit proposed by Illum *et al.* [2] is only marginally better than that for the plot in Figure 2. Given that the molecular weight of the hydrophobe cannot be expected to be a perfect correlation to the strength of binding, this degree of correlation indicates that alternative explanations are possible for (or perhaps that other factors can also contribute to explain) the results of the phagocytosis experiments. The need for a strong interaction between the hydrophobe of the surfactant and the solid particle has been demonstrated by the fact that poloxamers do not prevent opsonisation of biodegradable drug carriers (such as poly(lactic acid), see Carstensen *et al.* [4]). This is because certain carrier particles are not sufficiently hydrophobic to provide a driving force for poloxamer adsorption. Lukowski *et al.* [5] have attempted to produce uncoated polymeric spheres of different surface energy, in order to see whether the opsonisation process could be prevented without any steric stabilisation. As the acrylic acid content of the polymerised materials was increased, the contact angle (of water) measured on the surface decreased from 83 to 48° (due to the increasing numbers of hydrophilic carboxylic acids groups that were expressed in the surface). Unfortunately, none of the particles were sufficiently hydrophilic to prevent opsonisation, which is in keeping with the findings of other workers [6] who noted that above a critical contact angle value, the adsorption of fibrinogen was not significantly different for different surfaces.

Rudt and Muller [7] have investigated phagocytosis of polystyrene latex particles coated with each of 22 different poloxamers. Relationships were probed between the reduction in phagocytic uptake and the total surfactant molecular weight, and the poly(oxypropylene) molecular weight. Rudt and Muller [7] noted that there was a "tendency" for phagocytic uptake to reduce when the surfactant molecular weight was greater than approximately 4000. This is obviously a generalisation, from which there are a few exceptions, notably poloxamer 108 and 333 have identical molecular weights, but relative phagocytic uptakes of 94.5 and 3.3 % respectively. When considering poloxamers with the same poly(oxypropylene) chain length, variations in the length of the hydrophilic poly(oxyethylene) did not seem to have a great effect on phagocytic uptake. For example, the relative phagocytic uptakes of poloxamers 331, 333, 334, 335, and 338 were 22.9, 3.3, 2.5, 7.2 and 6.0 % respectively. The structures of these poloxamers are given in Table 1. The effect of variation in poly(oxypropylene) chain length is, however, seen to be significant. Poly(oxypropylene) chains with greater than 39 units seem to provide better protection against phagocytosis, which presumably must be due to more secure adsorption to the surface. The secure adsorption may well prevent displacement of the adsorbed poloxamer by other species, and thus limit phagocytosis. Part of the mechanism for this may be linked to steric stabilisation effects. However, if the ethylene oxide chains are too

Figure 2. A plot of % phagocytic uptake as a function of the coating layer thickness of poloxamer on polystyrene latex (data from Illum and Davis[2]).

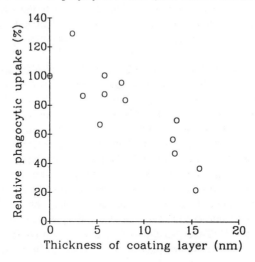

Figure 3 The data from Figure 2 replotted as % phagocytic uptake as a function of molecular weight of the poly(oxypropylene) hydrophobe of the poloxamers.

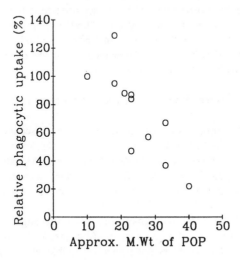

short, then inevitably the surface is not made sufficiently hydrophilic to prevent opsonisation, neither is there any possibility of steric stabilisation. It follows that for poloxamers with short ethylene oxide chains (3-7 units) the effect of varying the poly(oxypropylene) chain length is minimal. The effect of poly(oxypropylene) chain length is greatest if the hydrophilic chains are longer than 46 ethylene oxide units.

Porter *et al.* [8] have noted that different batches of poloxamer 407 (from three different suppliers) do not all give the same distribution of coated polystyrene latex throughout the body (rabbit). The earlier study with 407 [9] showed that the latex particles were directed to the bone marrow, but in the recent study [8] the 407 from Eugine Kulhman (UK) directed some 70% of the dose to the carcass, whilst 407 from ICI (France) and BASF (USA) directed approximately 40 and 30 % respectively, to the carcass, with corresponding increases in the deposition in the liver. The adsorption layer thickness for the coating did not correlate with the changes in body distribution, and consequently the idea that steric stabilisation is the only reason for the delivery to the bone marrow must be dismissed.   The three batches of material all had bimodal molecular weight distributions (which is quite normal for poloxamer 407), and all had different molecular weights. Porter *et al.* [8] "believe that the surface density and conformation of poloxamer 407 on the surface of the microspheres could play an important role" in surface recognition and distribution. Davis *et al.* [10] noted that if the steric stabilisation mechanism were effective, then particle - cell interactions should be prevented, and thus there would be no reason for different fates of particles which all have sufficiently large steric stabilisation barriers.   It is, for example, impossible to see any major physicochemical difference between polystyrene latex particles that have been coated with poloxamer 407 and poloxamine 908 (in terms of surface charge, coating layer thickness and hydrophobicity), however, those coated with 407 accumulate in the bone marrow, whilst those coated with the poloxamine 908 circulate in the blood [11]. The difference in fate of these particles must be related to the proteins that adsorb to their surfaces. Despite the fact that the surface appear to be identical, they clearly must be different if they facilitate adsorption of different proteins.

2.3 Issues concerning the physical state of the poloxamer prior to adsorption

The fact that the protection offered by a poloxamer seems to be related to its strength of adsorption raises important issues about the conformation of the surfactant prior to adsorption. Our own studies [12] have shown that the extent of adsorption of poloxamer onto hydrophobic substrates will be dependent upon the temperature of the experiment with respect to the phase transition temperature of the surfactant.

The phase behaviour of poloxamers has been investigated in dilute aqueous solution [13-16], and in more concentrated systems (e.g. refs [17,18]). It is clear that dilute aqueous solutions of the poloxamers undergo a reversible phase change, at a defined temperature (T_p). The T_p for this response is concentration dependent, with a higher T_p being seen with decreasing concentration [16] (although this is not clear from a comparison of all the data of Mitchard *et al.* [14] and Armstrong *et al.* [16]). It has been shown that the T_p, and thermodynamic parameters relating to this transition, correlate with the poly(oxypropylene) content of the surfactants, rather than with either total molecular weight or the poly(oxyethylene) content [14]. The involvement of the poly(oxypropylene) regions has also been identified by NMR studies [15], which revealed that the only region in the structure to show a change in properties at the T_p of the surfactant was the CH_3 group on the poly(oxypropylene). Recently [19] it has been argued that these phase transitions are in fact the onset of micellisation (a critical micelle temperature, for a given concentration) as a consequence of a highly temperature dependent micelle formation.

The adsorption isotherms of different poloxamers have been measured on various hydrophobic surfaces at a range of temperatures. The amount adsorbed at a selected equilibrium concentration of surfactant has been plotted as a function of the temperature (shown in Figure 4 for p338 on atovaquone). The choice of the equilibrium concentration used to construct Figure 4 did not affect the shape of the plot within the range of concentrations studied. It can be seen that the amount adsorbed at an equilibrium concentration of 5 (or 10) mg L^{-1} initially falls with temperature (in line with an exothermic event), but then rises before subsequently falling again. The peak for adsorption is in each case in the order of 8 degrees above the T_p reported by Mitchard *et al.*[14] and 6 degrees above the T_p values reported by Armstrong *et al.*[16] (who each used a different concentration of surfactant). The rise in T_p with a decrease in concentration is in keeping with the observation reported by Armstrong *et al.*[16] and with the observations

Figure 4.   Adsorption of poloxamer 338 to a model hydrophobic drug substance as a function of temperature. Phase transition for this polymer was reported as being 31.8 °C [14], but the transition can be expected to be concentration dependent. (Data of Carthew, Buckton, Parsons and Poole)

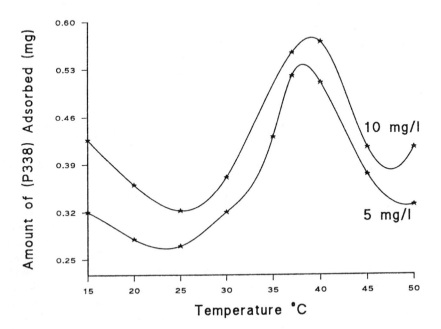

on temperature dependence of the critical micelle concentration reported by Alexandridis *et al.*[19]. It seems that the transition has great significance at the very low concentrations (5 mg L^{-1}) that have been used in this study (which are too dilute for calorimetric investigation).

The increase in adsorption noted at the phase transition is in keeping with the view that the hydrophobe of the surfactant has densified, thus making it possible for more surfactant to adsorb to the same surface area of solid. This model of behaviour is common, even if the "phase transition" is indeed due to micelle formation. It is likely that the interaction between the solid and the densified hydrophobe will be different to that between the solid and the surfactant below the phase transition, this may well relate to the functionality of the surface coating. For example, the fact that the organ distribution of poloxamer coated colloidal particles correlates with poly(oxypropylene) content, as does the T_p, may well mean that the proximity to the phase transition has an influence of the structure of the coated surface, which in turn has the significant influence on functionality. It is possible that many unexplained occurrences, such as batch to batch variability on organ distribution following adsorption of P407[8] are a consequence of adsorption at, or around, T_p.

The adsorption process which occurred, both above and below the T_p, was irreversible with respect to temperature (tested by dilution of the adsorption experiment which revealed no significant desorption, and by heating the sub-T_p adsorption experiment, after equilibrium had been reached, which showed no evidence of further adsorption, as assessed by the colorimetric assay). This means that it is the temperature at which adsorption occurs that determines structure, rather than the temperature at which the coated particles are stored and / or used. There are, however, still possibilities that the conformation of the exposed poly(oxyethylene) chains may vary as a function of temperature after adsorption. It has been suggested, for example, that the distance that the chains protrude from the surface will alter in relation to the temperature-solubility profile of the poly(oxyethylene)[20] (although this observation was for poly(oxyethylene) chains which had been grafted to a surface rather than from adsorbed poloxamer).

3. THE TOXICITY OF NON-IONIC SURFACTANT VESICLES

Non-ionic surfactant vesicles are structures which are stabilised by the addition of cholesterol, such that drug can be entrapped. The vesicles contain an inner aqueous core and a bilayer structure and are directly analogous to liposomes. These structures have been used widely in the cosmetic industry and have been investigated as drug delivery systems. A concern with any use of surfactants will always be the potential toxicity caused by the interaction of the surfactant with biological membranes. Ionic surfactants tend to be more toxic than non-ionic molecules. The interaction of surfactants with biological membranes has a range of advantageous and disadvantageous applications, ranging from the bactericidal properties of certain surfactants, through the use of these classes of molecules as penetration enhancers to aid drug passage across biological membranes, to the ultimate problem whereby the surfactant irreversibly damages membranes of the host such that it causes gross toxicity. The probability is that non-ionic surfactant vesicles will all prove toxic if chronically administered to man; however, for dermal or transdermal administration, there are clear advantages in using such vesicles.

The advantages include improvements in distribution and toxicity profiles of certain drugs, the possibility of improved stability of drugs and improved delivery giving better bioavailability. The surfactants are also known to alter membrane permeability and as such they may function as permeation enhancers when used on the skin. It is important, however, to understand the reasons for the toxicity of these vesicles.

3.1 Consideration of toxicity tests

Hofland *et al*[21] investigated the toxicity of non-ionic surfactant vesicles consisting of poly(oxyethylene) alkyl ethers and cholesterol. The vesicles were made from surfactants of differing alkyl chain (n) and poly(oxyethylene) head group (m) lengths which can be represented by the notation of C_nEO_m. Two different toxicity models were used to assess the relationships between the physico-chemical properties of the vesicles and their safety for topical administration, these were ciliary beat frequency (i.e. the retention of viability of nasal epithelia) and cell proliferation of human keratinocytes. The toxicity was found to be uncorrelated with the surfactant critical micelle concentration and also with the hydrophilic-lipophilic balance of the surfactants. There was, however, good correlation between the gel-liquid transitions of the surfactants and the vesicle toxicity. The vesicles were toxic if they were above their gel-liquid transition point and safe if below. This relationship showed direct proportionality between the transition point and toxicity in that surfactants which were tested close to their transition showed marginal toxicity, whilst those which were well below (e.g. 10°C below) showed no toxicity, with those which were well above showing extreme toxicity. It follows that the study of phase behaviour for these surfactants is extremely important.

3.2 Scanning calorimetry of the surfactants

The surfactants have been investigated as dilute dispersions in water using a Microcal MC-2 high sensitivity differential scanning calorimeter (HSDSC) (up to 15 mg mL^{-1} in 30mM phosphate buffered saline at pH 7.2 at 1 K min^{-1}) and in the solid state using a Perkin Elmer DSC7 (DSC) (2.5 mg load at 2.5 K min^{-1}) [22-23].

A comparison of the solid state DSC and HSDSC revealed certain similarities. Both the solid state and the dilute dispersion scans showed a pre and a main transition (which are shown for the HSDSC in Figures 5 and 6).

3.2.1 Pre-transitions. In both cases (solid state DSC and aqueous dispersions in HSDSC) the pre-transitions were identical. The pre-transition was seen to be due to many cooperating bodies and was related to a solid state structure which endured in the dispersed system. The T_m increased by ca 2 K per methylene group and decreased by ca 2/3 K per ethylene oxide group, indicating that the structure responsible was stabilised by being more hydrophobic. For further discussion and detail of the data, see Buckton *et al.*[23].

Figure 5.   HSDSC data for the main transitions of three different poly(oxyethylene alkyl ethers dispersed in water (3 mg mL^{-1}). Key: BC-2 = C$_{16}$EO$_2$; BS-2 = C$_{18}$EO$_2$; BS-4 = C$_{18}$EO$_4$. Reproduced from Buckton *et al.*[23], with permission.

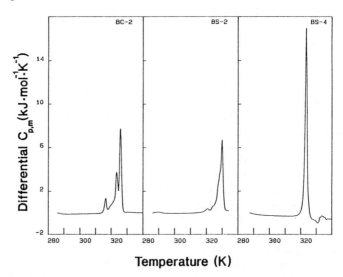

Figure 6.   HSDSC data for the pre-transitions of three different poly(oxyethylene alkyl ethers dispersed in water (3 mg mL^{-1}). Key: BC-2 = C$_{16}$EO$_2$; BS-2 = C$_{18}$EO$_2$; BS-4 = C$_{18}$EO$_4$. Reproduced from Buckton *et al.*[23], with permission.

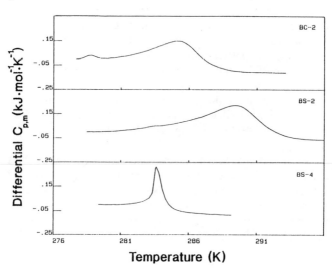

3.2.2 Main transitions. The HSDSC main transitions were significantly different to those seen in the solid state. For the HSDSC the T_m was always 40 K above the pre-transitions. The main transition T_m was always lower in the solid state than for the HSDSC, showing that water stabilises the structure. There were clear relationships between T_m and both the alkyl chain and poly(oxyethylene) head group lengths. The structure was stabilised by lowering the EO and increasing the alkyl chain lengths (i.e. making it more hydrophobic).

3.3 Relationship between surfactants and lipid bilayer transitions

There is a considerable amount of data on HSDSC of lipid bilayers. The surfactants studied here and lipid bilayers both exhibit a pre and a main transition. For the lipids the main transition is described as an alkane melt. Nagle[24] has shown that for a series of lecithins the T_m increases as a function of alkyl chain length, as is seen for these surfactants. The major difference between lipid bilayers and the surfactant structures investigated here is that the lipids have two hydrophobic chains attached to each polar head group, whereas the surfactants have just one. This is, however, probably unimportant as the separation between the chains on the same backbone in lipids is known to be about the same as that between neighbouring molecules[24].

It follows that there are very close similarities between the data for lipid bilayers and for surfactant phase transitions. The fact that the phase transitions also link to toxicity gives hope that studies of this nature can both correlate with and provide future explanations for the surfactant-lipid bilayer interactions which result in toxicity.

4 CONCLUSION

In this Chapter the importance of conformational changes of non-ionic surfactants have been considered in relation to the impact on adsorption properties and toxicity.

The change in conformation leading to a different interaction between poloxamer surfactants and hydrophobic solids in aqueous fluids can be related to the changes seen by HSDSC. It is reasonable to postulate that the impact of such changes in adsorption could be very significant in terms of biological fate of injected coated colloidal particles. There is much to be gained from combining studies on surfactant structure with biological investigations. A more detailed discussion of surface properties in relation to the biological fate of surfactant coated colloidal particles has been presented elsewhere[25]. The data for the poly(oxyethylene) alkyl ethers show a very clear correlation between HSDSC data and toxicity. A more detailed examination of the series reveals that the trends in the data show obvious structure-property relationships. Similar HSDSC transitions are seen for lipid bilayers. The prospect that surfactant toxicity can be understood through such studies is clear.

In general this work demonstrates the value of making appropriate physico-chemical measurements when considering interactions between surfactants and biological systems.

References

1. L. Illum and S. S. Davis, *FEBS Letters*, 1984, **167**, 79.
2. L. Illum, L. O. Jacobsen, R. H. Muller, E. Mak and S. S. Davis, *Biomaterials*, 1987, **8**, 113.
3. J. B. Kayes and D. A. Rawlins, *Colloid Polym. Sci.*, 1979, **257**, 622.
4. H. Carstensen, B. W. Muller and R. H. Muller, *Int. J. Pharm.*, 1991, **67**, 29.
5. G. Lukowski, R. H. Muller, B. W. Muller and M. Dittgen, *Int. J. Pharm.*, 1992, **84**, 23.
6. E. Brynda, N. A. Cepalova and M. Stol, *J. Biomed. Mater. Res.*, 1984, **18**, 685.
7. S. Rudt and R. H. Muller, *J. Cont. Rel.*, 1993, **25**, 51.
8. C. J. H. Porter, S. M. Moghimi, M. C. Davies, S. S. Davis and L. Illum, *Int. J. Pharm.*, 1992, **83**, 273.
9. S. S. Davis and L. Illum, In E. Tomlinson and S. S. Davis, (Eds.), Site Specific Drug Delivery, Wiley, Chichester, pp 93-110, 1986.
10. S. S. Davis, L. Illum, S. M. Moghimi, M. C. Davies, C. J. H. Porter, I. S. Muir, A. Brindley, N. M. Christy, M. E. Norman, P. Williams and S. E. Dunn, *J. Cont. Rel.*, 1993, **24**, 157.
11. R. H. Muller and K. H. Wallis, *Int. J. Pharm.*,1993, **89**, 25.
12. D. Carthew, G. Buckton, G. E. Parsons and S. Poole, *Pharm. Sci.*, 1995, **1**, 3.
13. N. Mitchard, A. Beezer, N. Rees, J. Mitchell, S. Leharne, B. Chowdhry and G. Buckton, *J. Chem. Soc., Chem. Commun.*, 1990, **13**, 900.
14. N. Mitchard, A. E. Beezer, J. C. Mitchell, J. K. Armstrong, B. Z. Chowdhry, S. Leharne and G. Buckton, *J. Phys. Chem.*, 1992, **96**, 9507.
15. A. E. Beezer, J. C. Mitchell, N. H. Rees, J. K. Armstrong, B. Z. Chowdhry, S. Leharne and G. Buckton, *J. Chem. Res. (S)*, 1991, **9**, 254.
16. J. Armstrong, J. Parsonage, B. Chowdhry, S. Leharne, J. Mitchell, A. Beezer, K. Lohner and P. Laggner, *J. Phys. Chem.*, 1993, **97**, 3904.
17. P. Linse, *J. Phys. Chem.*, 1993, **97**, 13896
18. G. Wanka, H. Hoffman and W. Ulbricht, *Macromolecules*, 1994, **27**, 4145.
19. P. Alexandridis, J. F. Holzwarth and T. A. Hatton, *Macromolecules*, 1994, **27**, 2414.
20. K. Holmberg, K. Bergstrom, C. Brink, E. Osterberg, F. Tiberg and J. M. Harris, In K. L. Mittal, (Ed.) Contact Angle, Wettability and Adhesion, VSP, Utrecht, pp813-827, 1993.
21. H. E. J. Hofland, J. A. Bouwstra, J. C. Verhoef, G. Buckton, B. Z. Chowdhry, M. Ponec and H. E. Junginger, *J. Pharm. Pharmacol.*, 1992, **44**, 287.
22. G. Buckton, B. Z. Chowdhry, J. K. Armstrong, S. A. Leharne, *Int. J. Pharm.*, 1992, **83**, 115.
23. G. Buckton, J. K. Armstrong, B. Z. Chowdhry, S. Leharne and A. E. Beezer, *Int. J. Pharm.*, 1994, **110**, 179.
24. J.F.Nagle *Annu. Rev. Phys. Chem.*, **31**, 157.
25. G. Buckton, Interfacial Phenomena in Drug Delivery and Targeting, Harwood, Amsterdam, 1995.

Monofunctional Poly(ethylene glycol): Characterisation and Purity for Protein-Modification Applications

M. Roberts and D. F. Scholes

POLYMER LABORATORIES, ESSEX ROAD, CHURCH STRETTON, SHROPSHIRE SY6 6AX, UK

1 INTRODUCTION

Historically, a major barrier to the use of novel peptides and proteins in medical therapeutics has been the effectiveness of the body's defense systems in recognising and removing 'foreign' materials from the blood stream. This defense system is primarily controlled by the scavenging properties of the phagocytic cells of the reticuloendothelial system located chiefly in the liver and spleen, but is aided by hepatic metabolism and the selective removal of small molecular weight molecules from the circulation by the renal glomerulus.

As such, biotechnology-derived proteins, particularly of low molecular weight, are easily recognised as foreign by the body and removed. Coupled with the elicitation of a specific immune response which speeds up clearance on subsequent exposure to the same molecule, the ultimate result is that the use of proteinaceous drug therapy often requires repeated, high doses in order to achieve the desired serum levels. This is particularly problematic in cases where the drug is expensive, has limited availability or severe side-effects.

Such complicated therapy has major cost and compliance implications and has been detrimental to the introduction of advanced therapies based on expensive biotechnology-derived proteins and peptides.

The discovery that the chemical attachment of poly(ethyleneglycol) (PEG) to proteins greatly reduced their antigenicity and immunogenicity while also dramatically increasing their serum half-life, all without adversely affecting their therapeutic efficacy, has obvious implications in the modification of such therapeutic molecules.

Historically, the term PEG has been used to describe all polymers of ethylene oxide with molecular weights of less than 20,000, irrespective of the nature of their end groups. Similarly, the contrasting term PEO has been used to describe the same materials of higher molecular weight. Within both categories, such blanket nomenclature has been used to describe polymers with either one or two hydroxyl end-groups. As the terminal hydroxyl groups form a ready site for covalent linkage in PEG-modification technologies, the control of the end-group characteristics is of paramount importance. For most applications requiring PEG-linkage, coupling is via a single hydroxyl group with it being desirable that the opposite end-group is chemically inert. Such PEGs have been offered commercially for many years, for example as the monomethoxy PEG possessing one methoxy and one hydroxyl functionality. It is such products which have aroused the largest interest in pegylation technology. In such 'mono-functional' preparations, the presence of 'diol' polymers (ie. polymer with two hydroxyl end-groups arising from poorly controlled polymerisation conditions) can cause major problems in terms of cross-linking of modified proteins with subsequent pharmacological and pathological implications. As such, it is important to accurately control each end-group and to be able to verify its nature. Unfortunately, until recently, users of supposedly monomethoxyPEG have needed to rely on the manufacturers purity data without an ability to confirm practically the claims. In such cases, suppliers have also assumed that their manufacturing techniques would produce pure monomethoxy materials without considering system impurities which may lead to the side formation of diol contaminants.

Early experiments in our laboratories demonstrated that in contrast to their accompanying technical data, many commercial sources of mPEG have significant levels of diol contamination and that the level of such contamination worsens as the molecular weight of the polymer increases. Until recently, a diol free polymer preparation of molecular weight greater than 10,000D was not available commercially. In light of the work of Sherman *et al* indicating that the modification of proteins with PEGs of higher molecular weights (up to 50,000 D) can produce significantly better results, this is somewhat disturbing.

Historically, users of high-diol products have purified them to remove the di-functional contaminant. Such additional down-stream processing of a fundamental component of many novel biotechnology products has caused great concern to regulatory bodies.

To satisfy increasingly stringent regulatory requirements it is essential to provide accurate, detailed characterisation of a material finding increasing application in advanced therapeutic formulations. The aim of this article, therefore, is to provide the user or potential user of monofunctional PEG with an overview of characterisation techniques applicable to the study of polymer molecular weight and diol content, both of which are known to have dramatic effects on the pharmacological profile of administered formulations. Such techniques are equally applicable to the manufacturer and the user wishing to establish an in-house characterisation facility.

2 PHYSICAL CHARACTERISATION

2.1 Polymer Molecular Weight and Molecular Uniformity

Molecular weight characterisation may be performed by a number of methods including gel permeation/size exclusion chromatography (GPC/SEC), light scattering and capillary viscometry. The industry standard method of polymer molecular weight characterisation is organic size exclusion GPC using suitable chromatography columns calibrated against defined polymer standards. Such analyses provide valuable information on average molecular weights and molecular weight distributions (polydispersity). A typical chromatogram of mono-functional PEG (Polymer Laboratories) illustrating the narrow, unimodal molecular weight distribution is given in Figure 1. The presence of a broad molecular weight distribution exemplified by a large polydispersity and / or multiple peaks in the spectrum would be indicative of a poor quality polymer which will produce poor, irreproducible results in protein-modification experiments.

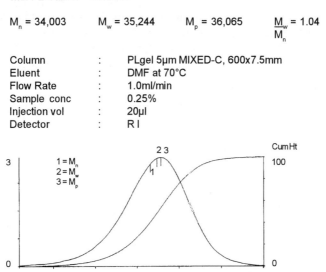

Characterisation using the PL GPC/SEC Chromatography Workstation with PL Caliber® Software

$M_n = 34,003$ $\quad$ $M_w = 35,244$ $\quad$ $M_p = 36,065$ $\quad$ $\dfrac{M_w}{M_n} = 1.04$

Column	:	PLgel 5µm MIXED-C, 600x7.5mm
Eluent	:	DMF at 70°C
Flow Rate	:	1.0ml/min
Sample conc	:	0.25%
Injection vol	:	20µl
Detector	:	R I

$1 = M_n$
$2 = M_w$
$3 = M_p$

Figure 1 *GPC / SEC Characterisation of a high quality PEG material to illustrate monodispersity*

3 CHEMICAL CHARACTERISATION

3.1 End-Group Determination

3.1.1 NMR Spectroscopy. A standard technique for polymer analysis is nuclear magnetic resonance spectroscopy (NMR) which can be a powerful tool for end-group characterisation, providing suitable chemical species exist in the end-groups to distinguish them from the polymer backbone. Although this may be easy for derivatised PEGs (see later) it is less straightforward for hydroxylated PEGs due to the uncharacteristic nature of this end-group. However if the hydroxyl end group(s) are chemically derivatised to a more 'NMR-friendly' species, the signals for each end-group become clearly distinguishable and it is possible to integrate the end-groups, determine the end-group ratio and thereby ascertain the percentage of diol in a given sample. For example, the method of Loccufier *et al*[1] describes reaction of the hydroxyl group with tcAIC which moves the hydroxyl protons downfield thereby making them clearly distinguishable.

A series of analyses were performed on monomethoxy PEGs of molecular weights between 2,000 and 10,000 obtained from a range of commercial sources and both monomethoxy and monobutoxy PEGs between 2,000 and 60,000 produced by the authors. Calculation of the end-group ratios of tcAIC derivatised PEGs of 2,000 MW in each case indicated that most of the polymers contained diol at a level between 2 and 4%. The polymer produced by the authors was diol-free. However, for commercial sources, as the molecular weight increased so did the diol content, with contamination at levels of up to 15% being seen for supposedly monomethoxy products of 10,000 MW.

Sample   :   PEG 20832-7(Polymer Laboratories)

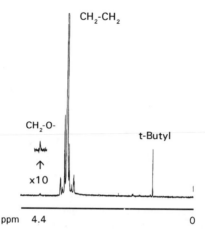

Figure 2 *NMR end-group analysis of a monohydroxy PEG to illustrate its diol-free nature*

In contrast, products produced by the authors did not suffer from a deterioration in quality at high molecular weights rather they provided exact 1:1 end-group ratios at molecular weight up to and in excess of 20,000 daltons. Although above this the end groups became more difficult to detect and extensive data accumulation runs were required, using a high-field NMR accumulating data for 24 hours, we were able to analyse a polymer of 30,000 MW. Figure 2 shows an NMR spectrum of a monobutoxy/monohydroxy PEG exhibiting characteristic peaks attributable to the polymer backbone and polymer end-groups. Even at such a high MW an end group ratio of 1:1.01 was obtained which is well within the constraints of the intrinsic error of the technique.

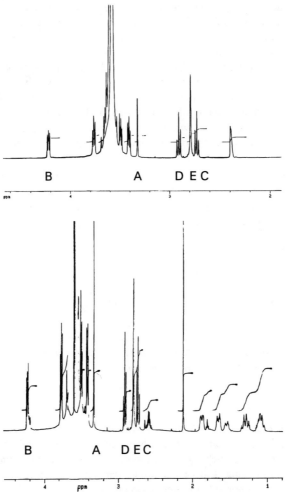

Figure 3 *Comparative NMR spectra of an ultra-pure, highly-active mPEG-succinimidyl succinate preparation and an impure, low-activity preparation of the same active nature*

To facilitate the attachment of drugs, the hydroxyl terminus of the PEG molecule is derivatised with a number of active species[2]. In our hands, high-field NMR has also proven to be a powerful tool in the determination of the level of substitution of such activated PEGs. NMR analyses allow not only the determination of the degree of substitution of the polymer molecule but also the identification (and quantitation) of process impurities. This is in contrast to standard biological assays which provide information only on the level of substitution. The lack of knowledge of the complete composition of such a material has obvious regulatory concerns.

Figure 3 contains spectra of two mPEGs activated with a succinimidyl succinate functionality, arguably the end-group most widely used in PEG-modification systems. The top spectrum contains each of the characteristic peaks attributable to the denoted chemical structure (Figure 4). Integration of peaks A and E shows the material to be 100% active and the remainder of the spectrum confirms the absence of any additional species which may be attributed to by-products such as mPEG-succinate, DCU and succinic anhydride. This is in contrast to the bottom spectrum which in addition to the fingerprint peaks seen above, contains additional peaks at < 2.7ppm due to these impurities. Further, integration of the end-group peaks for this polymer shows a reduced level of activation (75%) which can be attributed to non-optimal reaction conditions. Such information is of paramount importance to the end user.

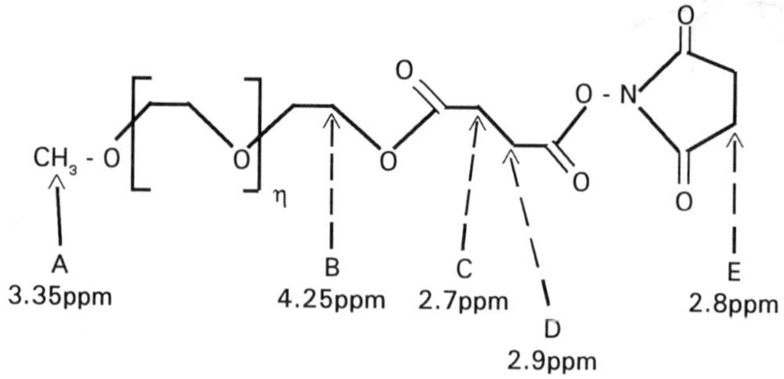

Figure 4 *Chemical structure of mPEG-succinimidyl succinate indicating the position of protons corresponding to those illustrated in Figure 3.*

3.1.2 Reversed-phase HPLC. Powerful as the NMR end-group analysis technique has proven to be, it has a significant disadvantage in that the end-groups become increasingly more difficult to integrate as the molecular weight of the polymer increases. In light of the previous comment that higher molecular weight polymers are becoming of primary interest, it was important to develop a simple test which would show the purity of polymers of molecular weights greater than 10,000 daltons. With the increasing emphasis on purity of the base mPEG material it is also important commercially to have a technique which facilitates rapid, reproducible analyses.

In a preparation containing both mono-hydroxy and di-hydroxy (diol) polymers, the two species differ slightly in polarity due to the differences in the nature of their end-groups. Using reversed-phase HPLC, it was possible to exploit these differences to effect the resolution of each species in a mixed population. Using a high purity monofunctional polymer which had been shown by NMR to be diol-free and a commercial 100% diol polymer of similar molecular weight as references, a calibration curve was obtained to assess the response of the detection system of choice (ideally an evaporative mass detector) against each material. By optimising the solvent and flow conditions it was possible to generate chromatograms with different retention times for the two individual polymers and for (w/w) mixtures of the two pure 'PEG' species. A change in the peak ratio was seen with variation in the mixture composition (Figure 5) and the correlation between the theoretical and experimental results with regard to percentage diol contamination was extremely good. Diol content data obtained for commercial samples using this method also correlated well with that obtained previously by NMR.

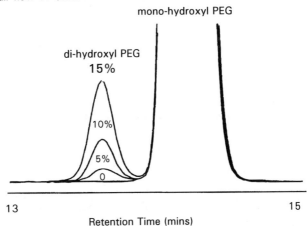

Column	:	PLRP-S 300Å 5µm, 150x4.6mm
Eluent A	:	99% Water / 1% Acetonitrile
Eluent B	:	100% Acetonitrile
Gradient	:	30-50% Eluent B in 20 mins
Flow Rate	:	0.5ml/min
Detector	:	PL-EMD 950 Evaporative Light Scattering Detector

Temp 75°C, air flow 11 l/min

mono-hydroxyl PEG

di-hydroxyl PEG
15%

10%

5%

0

13   15

Retention Time (mins)

Figure 5 *Chromatographic separation of mono- and di-hydroxyl PEG species in mixtures containing increasing percentages of di-hydroxyl (diol) PEG*

Reversed-phase HPLC has also been applied to the characterisation of activated PEGs where the introduction of an additional end-group species of significantly different polarity provides further improvements in separation efficiency.   The subsequent addition of a chromophore such as tryptophan onto the active end of the polymer also allows UV detection which may prove to be even more sensitive.

The combined techniques of high-field NMR and reversed-phase HPLC have proven invaluable as process quality control tools in the manufacture of ultra-high purity PEGs for pharmaceutical use.   They have provided important information on the chemical composition and purity of many commercial PEG preparations and allowed the acquisition of detailed profiles of the purity of a wide range of chemical functionalities. It is hoped that by expanding the techniques still further, it will be possible to develop assays suited to determination of the degree of substitution of a wide range of activated PEGs to allow the user to monitor the stability of the activated material during their particular process.

References

1.J.Loccufier, M.Van Bos and E.Schacht, *Polym. Bull.*, 1991, **27**, 201

2.J.M.Harris, 'Poly(ethylene glycol) Chemistry : Biotechnical and Biomedical Applications' Plenum Press, New York, 1992.

3.M.R.Sherman, L.W.Kwak, M.G.P.Saifer, L.D.Williams and J.J.Oppenheim, *Proc.Amm.Assoc.Conf. "Cytokines and Cytokine Receptors"* New York, October 1992.

Lactose – The Influence of Particle Size and Structure on Drug Delivery

H. J. Clyne

BORCULO WHEY PRODUCTS UK LTD, BRYMAU FOUR ESTATE, RIVER LANE, SALTNEY, NR. CHESTER CH4 8RQ, UK

1. INTRODUCTION.

Lactose used as an excipient has been around for many years and is still one of the most popular in use today for many pharmaceutical preparations. The reasons for its popularity are numerous, but the major ones are:-

- it is a natural product
- it is a stable material from a chemical, physical, and microbiological point of view
- it has a high compatibility with other excipients/actives.
- it is available in different forms.

These preparations may be in the form of tablets, capsules or powders. In general the lactose is present as a diluent or bulking agent, and in the past the particle size has been of minimal consideration, beyond whether it is to be a basic crystalline product, or in some generalised powdered form, the most well known example being the ubiquitous "200 Mesh".

Over the past few years changes in the manufacturing methods for both tablets and capsules, and the development of new dosage systems, have led to a re-appraisal of the role of the excipient, for two main reasons:-

(i) The introduction of high speed tableting and capsule filling machines has thrown the spotlight on the performance of excipients in respect to their powder properties. What will work happily on a machine producing 10,000 tablets per hours is suddenly a major efficiency problem when pushed to 100,000 per hour.

(ii) The effect of particle size on the delivery system of the active drug. These effects may be both direct and indirect depending on the particular role of the excipient within the delivery system. It has now been recognised that the so-called inert carrier can have major effects on the efficacy of a pharmaceutical preparation in areas such as rate of drug release.

PARTICLE SIZE DISTRIBUTIONS.

DIFFERENTIAL PLOT

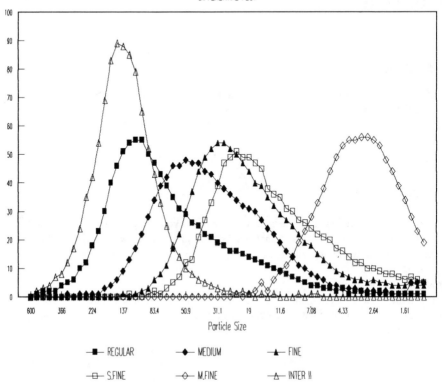

Particle Size

─■─ REGULAR   ─◆─ MEDIUM   ─▲─ FINE

─□─ S.FINE   ─◇─ M.FINE   ─△─ INTER II

2. PARTICLE SIZE DISTRIBUTIONS.

This graphic shows an example of the range of particle size distributions commercially available for lactose. This is by no means a definitive list, simply a depiction of the range. It is quite possible to tailor the distribution to meet an individual requirement.
This "tailoring" can be achieved in a number of ways. For example:

SIEVING.
MILLING.
AIR CLASSIFICATION.
MICRONISATION.
SPRAY DRYING.

or a combination of these processes.

3. PARTICLE SIZE TAILORING.

The use of powder processing techniques such as Milling, Air Classification, and Blending, in combination , can produce materials which are tailored in particle size distribution to specific needs.
In the first stage the raw crystalline powder is reduced to a specific mean size using a classifier mill. The resulting powder is then "split" into a coarse and fine fraction using an air classifier (see graphics on next page). This technique can handle fine powders that would be difficult, if not impossible, to sieve, due to their fine nature.

Finally the two fractions can be recombined in mathematically calculated combinations, to produce an product of precisely controlled distribution.

Lactose is being extensively used in formulations for inhaled dose forms. Here the aerodynamic properties of the excipient play a crucial role in delivering the drug to a selected area. These aerodynamic properties are directly affected by the particle size profile of the lactose.

Here precise particle size profiling can produce a narrow band distribution of fine powders using the Air Classification technique.

In this way the material can be controlled to achieve the necessary aerodynamic properties required in the carrier.

PARTICLE SIZE DISTRIBUTIONS.
DIFFERENTIAL PLOT

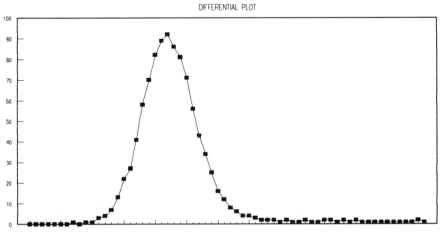

PARTICLE SIZE DISTRIBUTIONS.
DIFFERENTIAL PLOT

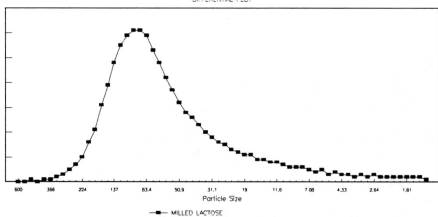

MILLED LACTOSE

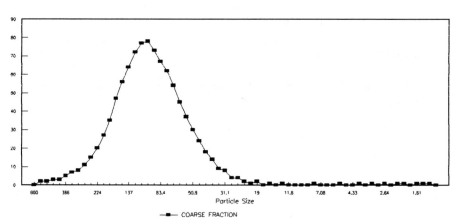

COARSE FRACTION

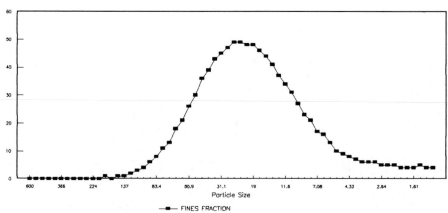

FINES FRACTION

4. POWDER PROPERTIES

The flow properties of the powder are very much a function of particle size distribution. This can have dramatic effects on the manufacturing of tablets. This has become particularly important in modern high speed machines

Particle size can also influence the dissolution rate of tablets, and so indirectly affect the rate in which drugs are released into the body.

The surface nature of the excipient can also have a major effect on the binding of drug entities. The reduction of particle size produces a corresponding increase in the surface area to volume ratio. This reaches an extreme form in Microfine Lactose which has a particle size mean below 5 microns.

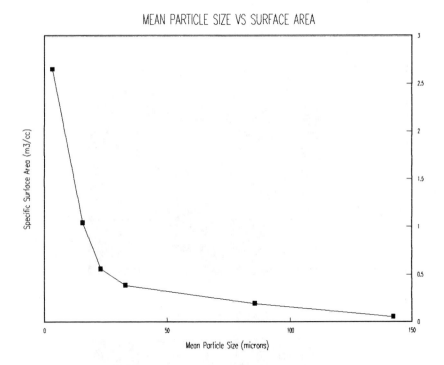

MEAN PARTICLE SIZE VS SURFACE AREA

This graph shows the relationship between the calculated surface area and the particle median diameter, as measured using laser diffraction. This calculation assumes that the particles are spherical and smooth. Unfortunately this is rarely the case in the real world.

5. PARTICLE STRUCTURE.

The surface characteristics of the excipient may have a major effect on how a drug entity interacts with the excipient.

The processing technique used in the manufacture of the powdered form of the excipient may therefore be an important consideration.

The electonmicrographs below show the effect of milling in the surface structure of a lactose crystal.

LACTOSE EM NO.1

LACTOSE E.M. NO.2

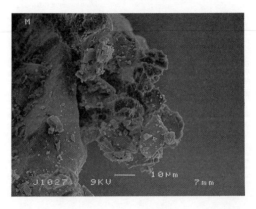

6. SPRAY DRIED LACTOSE

During production of spray dried lactose, modification of the particle size generates complex structures which are partially hollow in nature as seen in the electonmicrographs below.

This gives rise to a further potential to hold and bind drug entities. These particles achieve the mutually exclusive property of excellent powder flow, and rapid solubility, coupled with the ability to form tablets by direct compression.

ZEPAROX E.M. NO.1

ZEPAROX E.M. NO.2

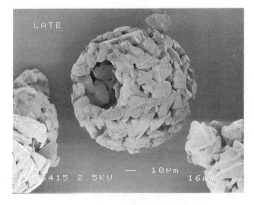

7. CONCLUSIONS.

Further, novel uses for lactose are being developed where the particle size is fundamental to its action and properties. This will require a more detailed study of the surface properties of the lactose crystal.

Functional Differences and Regulatory Aspects of Lactose Products Labelled as Lactose Modified

Jan J. Dijksterhuis

DMB INTERNATIONAL, UNIT PHARMA, PO BOX 13, 5460 BA VEGHEL, THE NETHERLANDS

1   INTRODUCTION

Novel manufacturing techniques for oral solid dosage forms become increasingly advanced and often also require more advanced raw materials than offered traditionally. In contrast to active materials where pharmacological properties and chemical purity are the main focus and interest, excipients are selected for their functional properties. These properties are not controlled by pharmacopoeial requirements but are subject to agreement between the manufacturer of the excipient and the user of the excipient.

Nevertheless in order to make identification possible lactose is labelled as "modified lactose" in case it is modified to its physical form and the method of modification should be indicated on the label or in the product description. However, the interpretation of what should be labelled as modified is subject for discussion because all lactose produced is physically modified in order to meet specifications for use.

1.1.   Manufacturing processes

Lactose is a naturally occurring carbohydrate and found in various concentrations in the milk of mammals. On an industrial scale lactose it is obtained by an isolation process from whey, the remaining liquid from a number of dairy products manufacturing processes. In the final process step lactose is obtained as crystals from a supersaturated and pure lactose solution but as such hardly suitable for use in pharmaceutical manufacturing processes. Although there are some methods to control particle shape and size during crystallisation (4), the product is not very well defined in terms of physical characteristics like particle size distribution and powder density. Furthermore the lactose crystals are of such a size that further processing is required to make the product suitable for use as an excipient.

In order to give the product the necessary functional properties as required for the different purposes in solid dosage form manufacturing, further processing is required. The processes used for further processing can be listed into three categories:

1.   Direct modification like
        - size reduction
        - size classification

2.   Advanced modification by secondary processing
        - particle shape modification
        - dehydration

3.   Co-processing of one or more different excipients

Examples of each of these three categories are:

Cat.	Products	Modification method	Intended effect
1.	Pharmatose®200M (DMV) Granulac (Meggle) Pharmatose®100M (DMV) Foremost 310 (Foremost Farms) Microfine lactose	Milling " Sieving " Micronisation	Granulation " Flow " Stabilisation
2.	Dilactose S (Freund) Pharmatose®DCL11 (DMV) Fast Flo 316 (Foremost Farms) Pharmatose®DCL15 (DMV) Tablettose (Meggle) Anhydrous alpha lactose	Granulation Spray drying " Granulation Granulation Thermal dehydration	Binding, flow " " " " "
3.	Ludipress (BASF) Cellactose (Meggle) Pharmatose®DCL40 (DMV)	Lactose,PVP,PVPP Lactose + cellulose Lactose + lactitol	Binding, disintegration, taste, flow, mixing

Table 1.

The regulatory status of products becomes different with increasing category number. The pharmacopoeial requirements primarily are based on identification requirements and chemical purity. Pharmacopoeial requirements seldom concern physical properties unless these can be used for identification purposes.

For the lactose products all products under category 1 do not need a specific labelling in order to discriminate. Although the various products in this class may be very different all are considered to be "regular" grades. That large differences may show between products is demonstrated in figure 1.

Here four different lactose monohydrate products with different particle sizes were used in the same granulation and processed under the same conditions into tablets. From the differences in tablet hardness obtained from the four different granulations it clearly shows that determination of particle size is necessary to control secondary

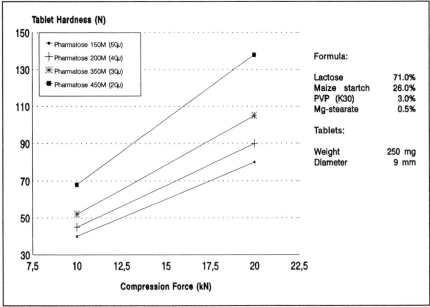

Fig.1: Effect of particle size on tablet strength

tablet parameters like hardness and porosity.

The second category are products which need to be labelled as "modified". This in practice means that a secondary processing step is carried out in order to tailor the physical characteristics to specified needs. For lactose this is primarily to enhance powder flow and compressibility in order to make the product suitable as a dry binder for direct compression or as a filler in combination with other dry binders. As demonstrated in figure 1 compressibility increases with decreasing particle size with translates in higher specific surface area (2). By incorporation of finely milled lactose particles in a granulate the enhanced compressibility is combined with good powder flow. Besides granulation the lactose can also be modified to its physical form. By using spray drying techniques part of the lactose can be transformed into an amorphous form which in contrary to the crystalline material demonstrates plastic flow under compression and consequently shows improved compactability (5).

The commercially available products in this second category all meet requirements for lactose as set out in the pharmacopoeia. Uncontrolled physical modification of lactose however may under certain circumstances result in failing the requirements as set out by the ruling monographs. An example for this is spray dried lactose which particles may be considered as agglomerates of fine lactose monohydrate crystals and amorphous lactose (6). Amorphous lactose does not contain water of crystallisation but due to its somewhat hygroscopic nature may absorb up to 15% water or, if stored under very dry conditions, will release significant amounts of water. Depending on the amount amorphous lactose present in the product the maximum or the minimum value for water as stated in the pharmacopoeias may not be met. Figure 2 demonstrates the different levels of moisture content for commercial and experimental spray dried products. All commercial available products do meet the requirements for lactose as per ruling pharmacopoeia but technological developments in this field are restricted by pharmacopoeial limits.

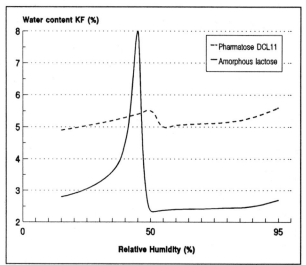

Fig. 2:   Water content (KF) at different relative humidity levels at 20°C

Lactose may contain the two enantiomers alpha and beta-lactose in variable amounts depending on the way the product is manufactured or processed. The alpha form can take up 5% water as water of crystallisation. The metastable beta form does not exist in a hydrous form. Depending on the composition of the final product the water content may vary from virtual zero to slightly over 5% for a pure alpha lactose monohydrate product. The pharmacopoeial limits restrict water content to be less than 1% in case of anhydrous lactose to in between 4.5 and 5.5% for hydrous lactose (1). In figure 3 it is depicted what the consequences of this are. Product development of new lactose based excipients will need to control the amount of beta lactose in order not to fail on one or both limits.

The Japanese Pharmacopoeia however amended the limits for water content to 4.0 to 5.5% since a local but widely used product Dilactose R, produced by Freund Industries contains a high level of beta lactose and consequently could not meet the required water content of 4.5%.

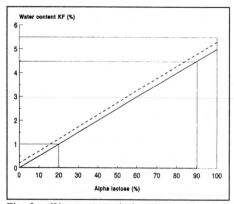

Fig. 3:   Water content in lactose products as a function of the amount of alpha lactose. The solid line represents the theoretical values. The dotted line includes adhered water.

1.2. Effect of processing on functional properties.

Although equal with respect to pharmacopoeial requirements, different lactose products may show incomparable results when applied in formulations. Products labelled as modified lactose and manufactured by the same method may still show significant differences in functional properties.

Table 2 lists the tablet characteristics of three equal formulations for lorazepanum 1 mg tablets. Figure 4 demonstrates the dissolution profile for the different tablets. All tablets were prepared on a rotary press by direct compression and the only difference was the type of DC lactose used. Significant differences in tablet hardness, disintegration and dissolution as well as in tablet strength and friability were observed. With all three lactose excipients it is possible to produce acceptable tablets but none of them is interchangeable. A precise description of the method of modification and characterisation on aspects relevant for the technical processing as well as an in depth knowledge of the product supplied by the manufacturer is an absolute prerequisite for use.

	Pharmatose DCL15	Pharmatose DCL11	Tablettose
Applied force (kN)	10.1	15.0	15.0
Binding capacity (N/kn)	6.4	4.3	3.0
Tablet hardness (N)	64.2	63.8	45.5
Tablet thickness (mm)	2.09	2.00	2.01
Friability (%)	0.09	0.06	0.11
Disintegration (sec)	88	367	18
Content uniformity (% s.d.)	3.0	2.3	5.5
Dissolution (min)	5	10	5

Table 2.: Lorazepam 1 mg tablets, 80 mg, prepared from the same formulation based on DC-lactose and manufactured on a Kilian rotary press.

1.3.   Co-processed products

Where the compendial issues are still not so much a problem for the modified lactose products, the class 3, co-processed products are not listed in any pharmacopoeia. The components individually may all meet the pharmacopoeial requirements or any other codex; the co-processed product does not conform with any monograph in the pharmacopoeia. Use of non compendial products is not common in pharmaceutical manufacturing and this status usually will prevent wide acceptance and use. Results of a questionnaire on the acceptance on non-compendial products are not promising. Figure 5 shows the attitude of senior US managers involved in dosage form formulation (3).

Consequently prod-

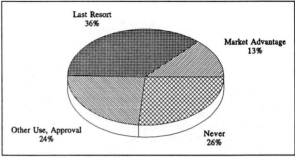

Fig. 5:   Non-compendial excipient policy for the major US pharmaceutical manufacturers. (After Shangraw c.s.)

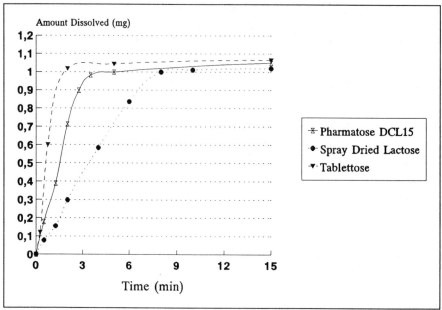

Fig. 4: Dissolution profile of Lorazepanum 1 mg tablets using 3 different DC-lactose products in a similar formulation.

uct development of better products with excellent functional properties is not stimulated because many manufacturers consider investments in development of new excipients as high risk in terms of returns. This situation is neither in the interest of the pharmaceutical industry, nor the manufacturers of excipients nor the consumers.

New excipients with improved properties could lead to improved tablet manufacturing techniques which may offer possibilities for efficiency improvements and consequently cost savings. In an era where the cost of manufacturing becomes more and more important initiatives are welcomed to overcome the problems related to regulatory status of excipients. At present innovative products like e.g. Pharmatose-®DCL40 and Cellactose may not be as widely used as just because of fear for delay in registration of new products.

2.   CONCLUSIONS

Pharmacopoeia monographs regulate the quality of products which can be used in pharmaceutical end products. In this way they act in the interest of the patient to safeguard for use ingredients not fulfilling requirements for quality. Because of the wide differences in tablet characteristics as demonstrated and as the result of differences in functional properties, the pharmacopoeia have started to discriminate products not exclusively on their chemical aspects. Although functionality of excipients cannot be described in unique terms because of the wide range of use, it is the manufacturer and end-users responsibility to agree upon and specify these. An in depth knowledge of the various functional aspects of excipients are needed to relate differences in tablet characteristics into excipient characteristics.

Development of generally accepted guide lines for description of excipients which are not listed in the pharmacopoeia would help to facilitate acceptance. A precise description in terms of composition, chemical and physical characteristics as well as the most important functional properties, a suggested assay and test procedures to check for composition and impurities profile would need to make part of the alternative monograph. (Semi-) governmental organisations like pharmacopoeia committees together with private organisations like JPEC/IPEC would jointly need to agree on the format of an excipient master file.

3.   LITERATURE

1.   Vromans H. c.s., Acta Pharm. Suec. 23 (1986) 231

2.   Vromans H. c.s., Pharm. Weekblad Sci. 7:186 (1985)

3.   Shagraw R.F., June 1992, Land-of-Lakes Conference

4.   York P., Drug Dev. Ind. Pharm. 18:677 (1992)

5.   Deurloo M. c.s., Pharm. Weekblad Sci. 14 Suppl.:F37 (1992)

6.   Kussendrager K. c.s., Acta Pharm. Suec. 18:94 (1981)

Flexibility in Tablet Formulation by Use of Lactose Based Direct Compression Compounds

Klaus Peter Aufmuth

MEGGLE GMBH, D-83512 WASSERBURG, GERMANY

1 INTRODUCTION

Direct compression (DC) of tablets by using DC-grade ingredients saves time and energy compared with multi-step operation procedures like wet granulation (WG). For an assessment of the overall cost savings by DC, a comparative study comprising four different tableting procedures has been carried out by a German pharmaceutical company.

Two DC-procedures, differing by the type of the mixer, have been compared with two WG-procedures based on fluidized bed granulation, one involving a separated mixing step. The main excipient of the placebo tablets was lactose, Tablettose 80 was used for DC. All four procedures have been carried out in a high and low cost excipients version respectively. Taking into account the three cost factors: labour, equipment and materials, it has been shown, that the price of the tablet mass produced by WG can be higher by more than 100 % compared with DC. The price difference becomes smaller with increasing batch size, where the batches are produced in campaigns, because no thorough cleaning is necessary between two sub-batches.

2 ADVANTAGES OF DIRECT COMPRESSION

DC has also distinct technological advantages over WG as follows:

- increased physical and chemical stability of tablets. Hardness and disintegration of tablets based on Tablettose 80 was almost unchanged over a storage period of three years, whereas wet granulated tablets showed a pronounced hardening with prolonged disintegration, particularly under dry storage conditions.

- more favourable dissolution profiles with higher initial dissolution rates. The dissolution value after 15 min of HCTH-tablets prepared by WG is approx. 15 % lower than the one of corresponding DC-tablets, based on published data.

Further advantages of DC are the better microbial purity of DC-excipients compared with the ones for WG, more narrow particle size distribution of DC-excipients than common WG-granulates as well as the higher humidity of the latter ones.

A frequent concern related to DC is the possible segregation of excipients and actives under vibration in the hopper of the tablet machine. It can be concluded from an investigation of the demixing stability of a 80 : 20 blend of Tablettose 80 with DC-grade ascorbic acid, that the blend quality is not adversely affected by the high vibration frequencies occuring in the tableting machine. Finely milled, low dosage actives are even better protected against demixing by the formation of interactive mixtures, which has been shown particularly for Cellactose, a spray-dried and macroporous DC-compound consisting of 75 % lactose and 25 % powdered cellulose. A blend of Cellactose with glibenclamide in various proportions retained more than 60 % of the active, suction resistant on a jet sieve due to physical adsorption (1). Lower adsorption values were obtained with sorbitol and MCC. Tablettose 70, an agglomerated DC-lactose with narrow particle size distribution, did not exhibit any demixing even after prolonged mixing with 1 % micronised prednisolone in a cube mixer, due to its well defined particle size distribution with reduced fines and coarse content. In admixture with 15 % of a 400 mesh lactose as a fine milled model ingredient the flow of Tablettose 80 through a funnel with 4 mm orifice was suppressed, whereas the flowability of Tablettose 70 was still acceptable with 0.5 g/sec in a similar trial.

3 TABLET FORMULATION WITH DC-COMPOUNDS

Compactibility and capacity (dilution potential) are important criteria for DC-excipients. In a study evaluating the formulation of low dose tablets based on various lactose and MCC types, tablets with the highest crushing strength, namely at equal compression force of 10 kN, were obtained using Cellactose 80 as filler/binder. Tablets produced by wet granulation and containing lactose and MCC in the same ratio as Cellactose showed a lower hardness of only 4.46 kg.

The hardness of low-dose antidiabetic tablets containing lactose and 28 mg Cellactose, total weight 170 mg, was 65.4 N, whereas a similar formulation with 28 mg MCC (Avicel PH 102) yielded tablets with 51.6 N hardness. The disintegration time was 290 sec and 400 sec respectively and the dissolution of the Cellactose containing tablets was somewhat retarded in the initial phase, possibly due to the higher hardness. During an accelerated stability test under various storage conditions, besides a loss of hardness from 65 N to 51-55 N, no significant change of the characteristics of the Cellactose containing tablets was observed, whereas MCC caused a drop in hardness from 51.6 N to 20 N (75 % rh/RT) accompanied by a decrease of disintegration from 290 sec to 40 sec and increase of friability from initially 0.59 % to 2.25 %. By use of super-disintegrants such as 2 % Croscarmellose, tablets having 80 sec disintegration time at 75 N hardness can be produced with Cellactose, with only slight increase to 120 sec at 125 N

The dilution potential of Cellactose is sufficient in order to produce medium dosage tablets without the need of further addition of a dry binder. For instance, Chlorthalidon tablets have been formulated with Cellactose according to the composition as follows:

Chlorthalidon Tablets 75 mg

Chlorthalidon	75.0 mg
Cellactose	220.5 mg
Aerosil 200	1.5 mg
Mg-stearate	3.0 mg
Total weight	300.0 mg

There is a nearly linear relationship between compression force and crushing strength of the tablets. No capping has been observed even at the high hardness values of 180 - 191 N. The compression ration of 0.91 is higher than the value 0.88 observed for coarse crystalline lactose, indicating a lower internal friction during the consolidation phase.

In high dosage formulations Cellactose can be used as sole filler-binder together with DC-grade actives or with extracts from medicinal plants. A tablet hardness of up to 120 N has been   reached with DC-grade vitamin C 98 % after addition of 22.5% Cellactose, sufficient even for chewing tablets. The capping point is at 25 kN compression force. Blends of Cellactose with Extr. Hippocastani containing maltodextrine as diluent can be compressed to tablets with various contents of extract up to 40 % without loss of hardness compared with the placebo tablets. The hardness yield with herbal extracts as actives depends on the diluent used by the extract manufacturer and its concentration in the extract.

Maltodextrin is a suitable diluent of extracts for tableting. For comparison tablets containing 10 % Extr. Hippocastani have been prepared using a blend of spray-dried lactose and MCC in the same ratio as Cellactose (75 : 25). The significant lower hardness yield obtained with MCC can be explained by the still considerable free water content of the aqueous extract, which exerts a stronger negative effect on the binding mechanism of MCC.

The mechanically interlocking fibrous cellulose component of Cellactose is less affected by residual water in actives. This is also the reason for the exponential correlation between crushing strength and disintegration time of tablets based, on Cellactose. Microcelac 100 (Meggle GmbH, FRG-Wasserburg) is a spray-dried DC-compound consisting of 75 % lactose monohydrate and 25 % microcrystalline cellulose type PH 101. Its cumulated mass flow through a funnel with orifice 4 mm after 70 sec reaches 57 g, that is 42 % more than the corresponding mass flow value of 75 parts of spray-dried lactose (Fast Flo 316) and 25 parts of MCC type 102 (Avicel PH 102). The reasons are the narrow particle size distribution, the spherical shape of the agglomerated articles and the strictly limited fines content of not more than 16 % passing through 230 mesh.

The hardness yield of Microcelac 100 is approx. 20 % higher than the one of Cellactose at equal compression force and approx. 50 % higher than the one of tablets made from lactose 100 mesh and 25 % MCC. By example of Furosemide tablets 40 mg it has been demonstrated that Microcelac 100 leads to a better dissolution rate in the initial phase than tablets made from spray-dried lactose and MCC. Microcelac 100 shows excellent mixing properties with micronized drugs due to its ability to incorporate the active in pores as recently known by Prof. Sunada, Meijo University, Nagoya (in print). When formulating enteric-coated tablets with polymethacrylate (Eudragit (R) L8OD), due to their very low friability, cores based on Microcelac 100 need approx. 30 % less coating material than cores produced from lactose and MCC in the ratio 75 : 25.

4 CONCLUSION

Lactose-based compounds for direct compression are flexible excipients for the formulation of low- and medium dosage tablets with various types of actives.

References

1.   P.C. Schmidt, Properties and Requirements for Tablet Excipients used on High Speed Rotary Tablet Presses. Proceedings Korsch Symposium, Berlin, 1993.

Compressional and Tableting Performance of High Density Grades of Microcrystalline Cellulose

G. E. Reier and T. A. Wheatley

PHARMACEUTICAL DIVISION, FMC CORPORATION, PRINCETON, NJ 08543, USA

1 INTRODUCTION

The advantages of the direct compression process of tablet manufacture are well known. The wide-spread applicability of the method on a large scale can be traced to the development in the early 1960's of one excipient - microcrystalline cellulose - and the modification of another - lactose. The modified lactoses are highly flowable but only moderately compressible by themselves and have no disintegrant properties of their own. Microcrystalline cellulose, on the other hand, is extremely compressible and possesses disintegrant properties. However, because of the average particle size (50μm) and the relatively low density of the initially available commercial material, Avicel® PH-101 microcrystalline cellulose, NF, its flowability was occasionally perceived to be a possible inhibitor to the flow of formulations, particularly during high speed tablet production. To address this concern, a larger particle size (90μm) microcrystalline cellulose was developed, Avicel® PH-102 microcrystalline cellulose, NF, which has about twice the mass flow of the original microcrystalline cellulose product and identical compressibility. An even larger particle size (190μm) microcrystalline cellulose, Avicel® PH-200 microcrystalline cellulose, NF, is available which also retains the compressibility of the original microcrystalline cellulose while having a still more rapid mass flow rate[1].

Nevertheless, there can be at times a need for microcrystalline cellulose products having the smaller particle sizes of Avicel® PH-101 and PH-102 microcrystalline celluloses but which have greater mass flow. Improvement in flow of these microcrystalline celluloses was accomplished by developing higher density products, designated as Avicel® PH-301 and PH-302 microcrystalline celluloses. The objective of this study was to investigate the compressional and tableting characteristics of these high density microcrystalline celluloses.

2 EXPERIMENTAL

The materials used in this study and a description of the experimental methods employed are as follows.

2.1 Materials

Ascorbic acid (Type S), USP, Hoffman-LaRoche, Inc. (Nutley, NJ); Avicel® PH-101 microcrystalline cellulose, NF, Ph. Eur., JP, BP, Avicel® PH-102 microcrystalline cellulose, NF, Ph Eur., JP, BP, Avicel® PH-301 microcrystalline cellulose, NF, Ph. Eur., JP, BP, and Avicel® PH-302 microcrystalline cellulose, NF, Ph. Eur., JP, BP, FMC Corporation (Philadelphia, PA); Cab-O-Sil®, M-5, colloidal silicon dioxide, NF, Cabot Corp. (Tuscola, IL); chlorpheniramine maleate, USP, Ashland Chemical Co. (Columbus, OH); Fast-Flo® Lactose, lactose, NF, Foremost Ingredient Group (Baraboo, WI); magnesium stearate, NF, Whittaker, Clark & Daniels, Inc. (South Plainfield, NJ); and stearic acid, NF, J. T. Baker, Inc. (Phillipsburg, NJ).

2.2 Methods

Loss on drying was determined by the USP procedure (<731>). A Scott Volumeter (Sargent-Welch Scientific Co., Skokie, IL) was used to determine loose density. Tapped density (600 taps) was determined using an Engelsman Tap Density Tester (J. Engelsman AG). Powder flow rates were determined using a proprietary device consisting of a 35° funnel holding 590 grams of material and having a 2 cm. diameter opening, to which is attached an air-driven vibrator. Particle size determinations were made using a Ro-Tap® apparatus and appropriate screens (W. S. Tyler Co., Cleveland, OH). Surface areas and total porosity were determined by the Quantachrome Corporation (Syosset, NY) using nitrogen adsorption and mercury intrusion techniques, respectively. Some scanning electron micrographs (SEMs) were obtained from Micron Incorporated Analytical Service Laboratory (Wilmington, DE); others were made using an Amray Model 1000B Scanning Electron Microscope (Amray, Inc., Bedford, MA).

Mixing was done using "V" blenders (Patterson-Kelley Co., East Stroudsburg, PA). Tablets were compressed on a 16 station Stokes B-2 rotary tablet machine (Stokes-Merrill Corp., Bristol, PA) instrumented by Specialty Measurements International (Whitehouse, NJ) so that compression force measurements could be obtained. The machine was operated at 42 RPM. Tooling was, unless otherwise noted, 7/16" round, standard concave. Four compression stations were used. A Manesty Express (Manesty Machines Ltd, Liverpool, England) was used only for the high speed compression evaluation (at 60 RPM) and all 20 compressing stations were used.

Tablet hardness was determined using a Schleuniger Model 6D hardness tester (Schleuniger Pharmatron, Inc., Manchester, NH). The method described in the USP (<1216>) was used to determine tablet friability. Disintegration in water was determined as described by the USP (<701>). For the determination of tablet thickness, an Ames digital gauge was used (Ames Corp., Waltham, MA).

The hardness/compression force profiles for each microcrystalline cellulose (previously passed through a 40 mesh screen) or microcrystalline cellulose/lubricant blend (both materials through a 40 mesh screen) were obtained by measuring the compression forces (40 replicates) obtained while compressing 400 mg at each of four target tablet hardnesses (4, 8, 12 and 16 Kp). Standard regression analysis was applied to the compression force data and actual tablet hardness data (10 replicates) determined after 24 hours. The results were plotted using hardnesses of 4, 8, 12, 16 Kp as the dependent variables (y axis) and the calculated compression forces as the independent variables (X axis).

The ascorbic acid dilution capacity index was determined by compressing 400 mg of mixtures of each microcrystalline cellulose with ascorbic acid, Cab-o-Sil® (0.5%) and stearic acid (2%). The amounts of ascorbic acid employed were 0, 20 and 35% with the amount of microcrystalline cellulose varying accordingly. Tablet compression was conducted at compression pressures beginning at 1000 Kg and increasing at 200 Kg increments up to 2400 Kg. Tablet hardness was determined after 24 hours, and hardness/compression force profiles constructed for each percent of ascorbic acid. Areas under the hardness/compression force profiles were integrated and further calculations applied using the method of Habib[2], *et.al.*, to yield the ascorbic acid dilution capacity index. This index is a measure of the amount of ascorbic acid that can be incorporated into the microcrystalline cellulose - ascorbic acid mix and still produce tablets, while taking into account the basic compressibility of the microcrystalline cellulose itself.

The content uniformity of the chlorpheniramine tablets was determined spectrophotmetrically at 262 nm following tablet disintegration and filtration. Dissolution was determined as described in the USP.

3 RESULTS AND DISCUSSION

3.1 Powder Properties

Powder property values of the high density grades of microcrystalline cellulose in comparison with those values found for Avicel® PH-101 and PH-102 microcrystalline cellulose are shown in Table 1. The moisture contents of all of the microcrystalline celluloses are well within the compendial specification of not more than 5%. PH-301 and PH-302 both exhibit about a 30% increase in loose and tapped density compared to PH-101 and PH-102. The mass flow of the lot of PH-301 tested was found to be about 35% greater than that of the PH-101 lot. The flow of PH-302 was about 15% greater than the flow of PH-102. The average particle sizes of the denser products as well as the percentages above a 100 mesh screen are comparable to those of their less-dense counterparts. The percentages through a 400 mesh screen are somewhat higher than those found for PH-101 and PH-102, thus indicating that perhaps PH-301 and PH-302 have

Table 1 *Powder Property Values*

	PH-101, Lot 1334	PH-102, Lot 2245	PH-301, Cork 2/94	PH-302, Cork 2/94
Moisture (% LOD)	3.8	3.6	2.8	2.9
Loose Density (g/ cm³)	0.31	0.31	0.41	0.40
Tapped Density (g/ cm³)	0.45	0.42	0.58	0.56
Mass Flow Rate (kg/min)	0.64	1.22	0.87	1.31
Average Particle Size (μm)	54	98	50	74
Percent above 100 mesh (150μm)	0.8	16.8	0.7	19.2
Percent below 400 mesh (38μm)	30.3	12.4	41.2	21.8
Surface Area (m²/g)	1.182	1.042	0.339	0.837
Total Porosity (cm³/g)	1.571	1.505	1.214	1.312

more fines and, to some extent, a wider particle size range, although not significantly so. This observation may well be batch related. The decrease in surface area and porosity found for the PH-301 and 302 lots compared to the PH-101 and PH-102 lots is to be expected given their increased density at about the same particle size. By definition, a material having a greater density, but approximately the same particle size as another, has less void space within it, and can have less external surface as well. The magnitude of the decrease in surface area of PH-301 compared to PH-101 (about 70%) is, however, striking.

Photomicrographs of PH-101, PH-102, PH-301 and PH-302 are shown in SEMs 1 to 4. The particle morphology for Avicel® PH-101 microcrystalline cellulose is identical to that of PH-301. There are individual microcrystals and microcrystal aggregates present. Typically, these are seen as log-like particles and irregularly shaped particles (aggregates). The particle morphology for PH-102 and PH-302 is also identical; however, because of the magnification and the particular fields chosen, the more dense nature of PH-302 as compared to PH-102 can be seen in their particular photomicrographs.

In order to determine whether the aggregates could be broken down into individual crystals, all four microcrystalline cellulose products were placed into a V-blender equipped with an intensifier bar and mixed for 5 minutes with the intensifier bar running. All four products showed a 4-6% decrease in average particle size which does not represent disintegration of aggregates, but rather probably a deaggregation. Previous experience has indicated that PH-101 and PH-102 average particle size is not decreased by even more severe shearing or milling conditions.

3.2 Tablet Weight Variation

Unlubricated material was passed through a 40 mesh screen and compressed with the results shown in Table 2. The more uniform flow of PH-301 and PH-302 as compared to PH-101 and PH-102 is shown by the tablet weight variation data. Tablets of PH-301 and PH-302 have lower coefficients of weight variation than do those compressed from PH-101 and PH-102 and are nearly identical within each study. PH-102 shows lower weight variation than does PH-101, consistent with observations that have previously been reported[1,3]. The values in Table 2 for the Express® are all the more remarkable given the fact that the materials were gravity fed to a standard (non-paddle equipped) feed-frame.

Table 2 *Tablet Weight Variation (% CV), Unlubricated Microcrystalline Cellulose*

	PH-101, Lot 1334	PH-102, Lot 2245	PH-301, Cork 2/94	PH-302, Cork 2/94
B-2 Compressing machine, @ 42 RPM, 4 stations, 7/16" round standard concave tools	4.9	3.4	1.3	0.8
	Lot 1423	Lot 2421	Lot P416C	Cork 8/94
Express® Compressing machine, @ 60 RPM, 20 stations, 3/8" round modified concave tools	19.1	7.2	3.0	3.2

3.3 Compression Properties

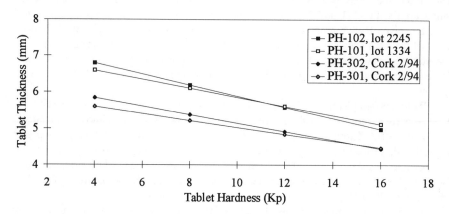

Figure 1 *Tablet Thickness/Hardness Profiles (Unlubricated)*

As would be anticipated from their increased density, unlubricated tablets compressed from PH-301 and PH-302 are thinner than tablets compressed from PH-101 and PH-102 at an equal weight (400 mg), size and hardness, as shown in Figure 1. Interestingly, the average decrease in tablet thickness when comparing tablets of PH-101 to PH-301 (about 14%) is similar to that when comparing tablets of PH-102 to PH-302 (about 12%).

The hardness/compression force profiles of unlubricated PH-101, PH-102, PH-301 and PH-302 are shown in Figure 2. The compressibilities of PH-101 and PH-102, as defined by the lines in the figure, are equal as has previously been reported[1,4]. The compressibilities of PH-301 and PH-302 are not quite equal, but they do not appear to be significantly different and probably should be considered equal considering the lines have the same slope. However, the compressibilities of both are less than that of PH-101 or PH-102. This is to be expected based on the increased density of the particles of PH-301 and PH-302. Particles having the same particle size but containing more mass within them cannot compact under equal compression pressures to the same degree as

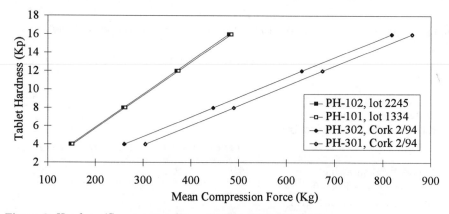

Figure 2 *Hardness/Compression Force Profiles (Unlubricated)*

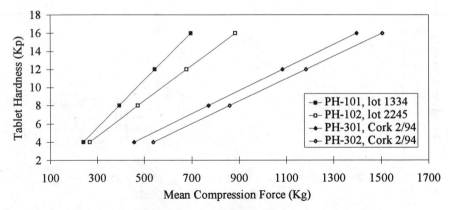

Figure 3 *Hardness/Compression Force Profiles (0.5% Magnesium Stearate)*

particles that are less dense. The effect of microcrystalline cellulose density on compactability is further illustrated by the fact that Avicel® PH-200 microcrystalline cellulose while approximately twice as large in particle size as PH-102, and four times as large as PH-101, but having approximately the same density, is equal in compressibilty to PH-101 and PH-102[1].

The data plotted in Figure 3 show that PH-101 and PH-102 lubricated with 0.5% magnesium stearate no longer have superimposable compressibility. The lubricant affects PH-102 to a greater degree than PH-101 as can be seen by the obviously lower slope of the PH-102 line. This observation is consistent with previous observations in regard to the greater sensitivity of PH-102 to magnesium stearate[1,5] and probably has to do with the fact that PH-102 particles are twice the size of PH-101 particles. There is less surface area in an equal weight of the former than there is in the latter and the magnesium stearate can then form a thicker coating around the particles of PH-102 and interfere with bonding to a greater extent then it does in the case of PH-101. This explanation can also be applied to the observation that PH-200, with its still larger particle size, is affected by magnesium stearate even more than is PH-102[1,5]. The hardness/compression force lines for PH-301 and PH-302 are parallel with equal slopes. Their lesser compressibility compared to PH-101 and PH-102 also is evident. Even though the surface area for PH-302 reported in Table 1 is greater than that for PH-301, the increased sensitivity to magnesium stearate would indicate otherwise.

Figure 4 shows the lesser compressibility of PH-301 and PH-302 compared to PH-101 and PH-102 when lubricated with 2% stearic acid. In the case of PH-101 and PH-102, the lines are nearly parallel but separated with PH-101 being somewhat less compressible. This reversal in lubricant sensitivity compared to magnesium stearate may be based on the difference in compressibility between stearic acid and magnesium stearate. Stearic acid alone can be compressed into a waxy solid, whereas magnesium stearate cannot. PH-102, with its larger particle size and lower surface area, again has a thicker lubricant layer than does PH-101. However, stearic acid in its thicker layer on PH-102 is compactable to some extent while the thinner layer on PH-101 is too diffuse to form a continuous film and serves only to interrupt bonding of the microcrystalline cellulose particles. Thus, the latter is not as compressible as the former. The relative positions of PH-301 and PH-302 are the same as their less dense counterparts

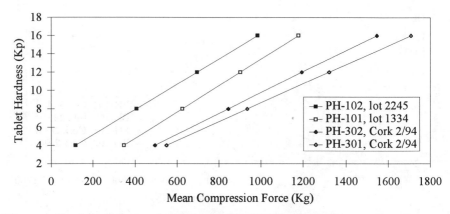

Figure 4 *Hardness/Compression Force Profiles (2% Stearic Acid*

in Figure 4 possibly also because of the thickness of the stearic acid layer on the particles.

The slopes of the hardness/compression force profiles given in Figures 1- 4 are shown in Table 3. The slope of a hardness/compression force plot is typically used to assign a value to the compressibility of a material. A decrease in compressibility caused by the addition of a lubricant, or other material, is manifested by a lower slope value. Comparisons of slope values are used to quantitate differences in the effect of various lubricants on compressibility (lubricant sensitivity). Also included in Table 3 are the percentage decreases in slope that were observed for the lubricated materials.

From the percent decrease it can be concluded that PH-301 and PH-302 are more sensitive to 0.5% magnesium stearate than is PH-101, but their degree of sensitivity is approximately equal to that of PH-102. Both PH-301 and PH-302 are less sensitive to 2% stearic acid than are PH-101 and PH-102. However, the magnitude of the decrease in compressibility of all the microcrystalline celluloses in the presence of 2% stearic acid was surprising.

Caution in data interpretation must be exercised when relying on hardness/compression force slopes alone. For example, the PH-301 and PH-302 compression force/hardness profile lines in Figure 2 are parallel lines and, of course, have the same slope, but somwhat higher compression forces are required to compress PH-301 than PH-302 to the same hardness. Further, the slopes of all the lines in Figure 4, while different, are numerically similar. However, the numerical value of the slope for PH-101

Table 3 *Slopes From Hardness/Compression Force Profiles (x10⁻³)*

	Unlubricated	0.5% Magnesium Stearate	2% Stearic Acid
PH-101	36.1	26.2 (27%↓)	14.5 (60%↓)
PH-102	36.3	19.4 (47%↓)	13.8 (62%↓)
PH-301	21.6	12.7 (41%↓)	10.4 (52%↓)
PH-302	21.6	12.4 (43%↓)	11.4 (47%↓)

↓ = % decrease from unlubricated value.

Table 4 *Ascorbic Acid Dilution Capacity Indices*

PH-101, lot 1334	17.9
PH-102, lot 2245	17.6
PH-301, lot P416C	14.7
PH-302, Cork 8/94	8.1

exceeds that for PH-102 while the relative position of the lines indicates that PH-102 is the more compressible of the two materials.

3.4 Ascorbic Acid Dilution Capacity Index

The ascorbic acid dilution capacity indices for the microcrystalline celluloses evaluated are given in Table 4. The results indicate, as would be expected based on the compressibility of the neat materials and a consideration of their physical form, that the denser grades of microcrystalline cellulose have a lesser ability to carry non-compressible materials. The fact that the lots of PH-101 and PH-102 examined exhibited an equal ascorbic acid dilution capacity index is to be expected. The differences found between the indices for PH-301 and PH-302 are not explainable.

3.5 Model Formulation Evaluation

The comparative properties of a model tablet formulation containing the four microcrystalline cellulose grades compressed using 5/16" round standard concave tooling are given in Table 5.

When the denser grades of microcrystalline cellulose were used, the tablet weight and hardness variations were reduced, particularly in the case of PH-302. The use of the higher density materials resulted in thinner tablets. Higher compression forces were required to compress the tablets containing PH-301 and PH-302 as one would expect from their neat hardness/compression force profiles. However, these differences in compression

Table 5 *Model Formulation Tablet Characteristics*

Characteristic	PH-101 lot 1334	PH-102 lot 2245	PH-301 Cork 2/94	PH-302 Cork 2/94
Weight Uniformity (% CV)	1.4	1.5	1.3	0.9
Hardness (Kp)	6.5 ± 0.6	6.6 ± 0.6	6.8 ± 0.5	6.6 ± 0.3
Thickness (mm)	3.60	3.58	3.56	3.49
Mean Compression Force(Kg)	676	669	782	744
Friability (%)	0.2	0.2	0.2	0.2
Content Uniformity (% CV)	4.4	1.3	1.4	1.3
Disintegration Time (sec.)	125	139	144	149
% Drug Dissolved in 15 min.	81	85	86	82

Formulation: Chlorpheniramine Maleate, 4.00 mg; Fast-Flo® Lactose, 122.65 mg; Microcrystalline Cellulose, 42.59 mg; Stearic Acid, 0.85 mg.

forces are not considered significant. The tablet friabilities did not change upon substitution of PH-301/PH-302 for PH-101/PH-102. The tablets containing PH-101 exhibited the highest variation in content uniformity. Additional experience is needed with other active ingredients in other formulations to determine if this particular observation would always be the case. Tablet disintegration time was higher for the products containing PH-301 and PH-302, but the disintegration property associated with microcrystalline cellulose obviously remains largely unaffected. Drug dissolution was approximately equal for all the products tested and far exceeds the USP dissolution specification for chlorpheniramine maleate tablets, which is not less than 75% dissolved in 45 minutes.

4 CONCLUSIONS

Avicel® PH-301 microcrystalline cellulose and Avicel® PH-302 microcrystalline cellulose are products that have higher loose (and tapped) densities and mass flow rates than Avicel® PH-101 microcrystalline cellulose and Avicel® PH-102 microcrystalline cellulose while maintaining the same average particle size. Because their particle size remains unchanged compared to PH-101 and PH-102, these new products have less internal porosity and, probably because of the decrease in porosity, less compressibility than either PH-101 or PH-102. This decreased compressibility made no practical difference in functional properties of a model tablet formulation. The decreased density of PH-301 and PH-302 does indicate, depending on the particular formulation, that thinner, or otherwise smaller dimensioned tablets, could be obtained. Content uniformity of tablets might be improved by using these high density products, although more data are needed. Indications are that weight uniformity is improved. The increase in density and consequent increase in mass flow of these products could result in increased flow rates of a formulation which, if true, could increase the rate of tablet production while maintaining present levels of tablet quality.

Acknowledgment

The authors gratefully acknowledge the contribution of L. M. DiMemmo in providing the scanning electron micrographs.

References

1. Avicel® PH Application Bulletin #26-2, FMC Corporation, Philadelphia, PA 19103.
2. Y. Habib, L. Augsburger, G. Reier, T. Wheatley, R. Shangraw, Poster presentation at AAPS Tenth Annual Meeting, Miami Beach, FL, November, 1995.
3. E. Doelker, *Drug Dev. Ind. Pharm.*, 1993, **19**, 2399.
4. J. W. Wallace, J. T. Capozzi, R. F. Shangraw, *Pharm. Tech.*, 1983, 7 (9), 94.
5. E. Doelker, D. Mordier, H. Iten, P. Humbert-Droz, *Drug Dev. Ind. Pharm.*, 1987, **13**, 1847.

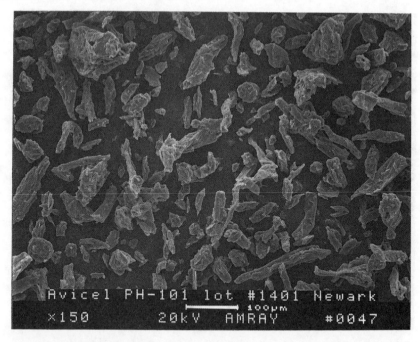

SEM 1 *SEM of Avicel® PH-101 microcrystalline cellulose, lot 1401*

SEM 2 *SEM of Avicel® PH-301 microcrystalline cellulose, P416C*

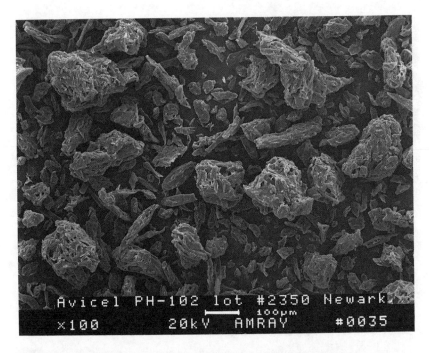

SEM 3 *SEM of Avicel® PH-102 microcrystalline cellulose, lot 2350*

SEM 4 *SEM of Avicel® PH-302, microcrystalline cellulose, lot Cork 8/94*

Starch Based Drug Delivery Systems

J. P. Remon[1], J. Voorspoels[1], M. Radeloff[2] and R. H. F. Beck[3]

[1] LABORATORY OF PHARMACEUTICAL TECHNOLOGY, UNIVERSITY OF GENT, BELGIUM
[2] CERESTAR INTERNATIONAL SALES, AVENUE LOUISE 129, BTE.13, B-1050 BRUSSELS, BELGIUM
[3] CERESTAR APPLICATION CENTRE PHARMA & CHEMICAL, HAVENSTRAAT 84, B-1800 VILVOORDE, BELGIUM

ABSTRACT

Starch is one of the most widely used excipients in the manufacture of solid dosage forms. It is used in conventional dosage forms as a filler, disintegrant or binder.

Starch has also been intensively studied as a basic material for drug delivery systems.

The paper reviews some of the more recent applications of starch and modified starches as oral sustained release matrix, bioadhesive microspheres, magnetic starch microspheres, erodible implants, microcapsules, bioadhesive formulations and as drug carriers.

1 INTRODUCTION

Besides the traditional and well established application of starch as diluent, binder and tablet disintegrant in solid dosage forms, the focus of this presentation is on reviewing the *more active* application potential of starch, whereby the molecular structure and characteristics of starch or modified starches are exploited for the delivery of drugs. In particular, after a short introduction on starch as a natural product, an overview on the use of starch in conventional dosage forms, as sustained release agent, as base material for microspheres and microcapsules, and as bioadhesive and drug carrier will be given.

2 STARCH - BASIC CHARACTERISTICS

Starch is the most abundant reserve polysaccharide of the plant kingdom, occurring in the form of small granules, ranging from about 1 to 100 μm in diameter. Starch can be found in nearly all organs of most higher plants, including pollen, leaves, stems, rhizomes, bulbs and fruits. However, the main storage areas of starch granules are seeds, roots and tubers. Starch is a heterogeneous material consisting of two types of polymers, amylose and amylopectin. The two polymers are structurally different: amylose is essentially linear and amylopectin has a branched structure. The ratio of amylose and amylopectin found in various starches varies depending upon the source. Most starches such as maize, wheat and potato contain 18 - 28 % amylose. Certain starches, for instance special hybrids of maize, such as waxy maize, are composed of essentially amylopectin. Basic molecular parameters such as the molecular weight distribution and crystallinity, which are reflected in the viscosity profile, the gelatinisation behaviour and the glueing characteristics of a starch, vary with the plant variety, the climate, the harvesting date, and storage condition of the crop. State of the art starch processing allows to correct these natural variances and guarantees for the customer a constant quality of starch raw materials.

3 STARCH IN CONVENTIONAL DOSAGE FORMS

Starch is one of the most widely used exipients in the manufacture of solid dosage forms and can be used as a filler, a disintegrant or as a binder. Several types of modified starch have been used as tablet excipients (e.g. Starch 1500® a physically modified partially hydrolysed starch, starches with a different amylose - amylopectin ratio, oxidized starches).

Starch is still the most common used disintegrating agent today. But because of the relatively large amounts of starch required and the its lack of compressibility, modified starches were developed and became popular, such as sodium starch glycolate, a low-substituted derivative of potato starch. Starch has also always been one of the most commonly used granulating agents and is usually used in the form of a paste. The use of cold water soluble (also sometimes referred to as *drum-dried, roll-dried, cold water swellable* or *pregelatinised*) waxy maize starch improves the binding properties. Worthwhile mentioning is the use of cold water soluble hydroxypropyl waxy maize starch. The main advantage is the ease of dispersion in preparing a paste; added in the dry form no difference in granule quality was observed in comparison with the paste addition.

Research has also been performed in order to use starch for reducing the quantity of gelatin needed for capsule for the production of hard gelatine capsules. The suggested material is a product of waxy maize starch and is a dextrin, chemically or thermally modified using succinic acid in the former case. Another modified starch used for this purpose is hydroxyalkyl starch.

4 SUSTAINED-RELEASE AGENT FOR ORAL DRUG DELIVERY

As a hydrophilic matrix, starch has received recently considerable attention. Several investigators[2,3] have mentioned the possible use of thermally modified starches for prolonged-release purposes. Different types of cold water soluble starches were evaluated as a hydrophilic matrix. The influence of the amylose-amylopectin content and of the technique and degree of pregelatinization on the final products were evaluated. The "cold water soluble" maize starch matrix showed sustained plasma concentration profiles similar to those of a commercially available product in humans. Te Wierik[4] in his work on the application and preparation of linear dextrins showed that amylodextrin and the metastable amylodextrins and metastable amylose produced tablets with excellent pressure-hardness profiles. Besides he reported on the successful application of amylodextrin as an excipient in the design of programmed release tablets. The release of active substances from both non-porous amylodextrin and metastable amylose tablets is governed by a mechanism of relaxation/controlled penetration of the solvent front in the tablet.

5 POLYETHYLENE/STARCH EXTRUDATES AS ERODIBLE CARRIERS

Krishnan et al[9] evaluated polyethylene-starch based carriers for sustaining the release of bioactive materials. Polyethylene-starch carriers were prepared by incorporating various amounts of maize starch 0, 40, 80 and 100 % in P.E. beads with and without dye. The granulated mixture was extruded. The dye release studies showed that the release could be sustained well over 12 weeks, depending on the percent starch incorporated. Scanning electron microscopy (SEM) showed gradual erosion of starch particles, leaving a polyethylene skeleton.

Lenaerts et al[26] introduced epichlorohydrin cross-linked amylose as a matrix for controlled release of drugs. A linear release of theophylline from tablets was observed in all cases. A linear increase of the cross-linking degree of the amylose used for tablet preparation generated a non-linear decrease in release time. Analysis of the data indicated that an anomalous release mechanism controlled the drug transport. A hypothesis of a release mechanism controlled by hydrogen association was proposed.

6 BIOADHESIVE MICROSPHERES

Illum et al[5] reported on bioadhesive microspheres that could not be cleared easily from the nasal cavity. In that system, albumin, starch and DEAE-dextran could be used as the formulation base. Studies have been conducted in human

volunteers using a variety of systems containing microspheres with good
bioadhesive properties. The microspheres prepared from starch and DEAE-
Sephadex were found to be effective in delaying nasal mucociliary clearance. The
half-life of clearance for starch microspheres was in the order of 240 min. as
compared to 15 min for the control formulations. Illum et al[5] investigated the
possibility of improving the bioavailability of gentamycin administered
intranasally by means of starch microspheres. If the gentamycin was administered
in combination with the starch microspheres, a significant increase in
bioavailability was obtained. An even more dramatic increase was seen when the
gentamycin/starch microspheres were administered together with
lysophosphatitdylcholine (Fig 1).

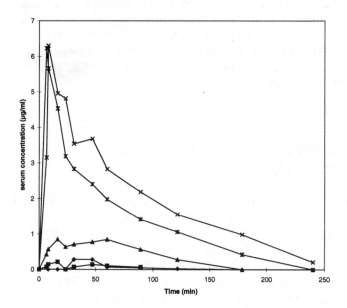

Figure 1 Comparative serum concentration profiles (meanvalues)
of gentamycin to sheep following intranasal administration of
(◆) gentamycin solution (5 mg/kg)(n = 2; (■) gentamycin solution
(5 mg/kg) containing 2 mg/ml of lysophosphatidylcholine (0.026
mg/kg)(n = 3); (▲) gentamycin (5 mg/kg) in starch microspheres
(n = 3); (✕) intranasal gentamycin (5 mg/kg) in starch microspheres with lysophosphatidylcholine (0.2 mg/kg)(n = 3) and
intravenous administration (✳) of gentamysin solution (2 mg/kg)
(n = 3).

The bioavailability was increased to about the same as that of the I.V. dose. Björk
and Edman[6] investigated a nasal delivery system for insulin in rats which used
degradable starch microspheres (DSM). Administered nasally as a powder, their

preparation gave a dose-dependent reduction in blood glucose and a concomitant increase in serum insulin. Their results indicated that the DSM system offers a means for improving the nasal absorption of drugs. Farraj et al[7] reported that using starch microspheres bioadhesive systems the half-time clearance of insulin from the nasal cavity was prolonged to 240 min in comparison to 15 min for a solution. In sheep the bioavailability of insulin was increased with ± 500 % when administered together with microspheres in comparison with an insulin solution.

7 MAGNETIC STARCH MICROSPHERES

Fahlvik et al[8] reported on the use of magnetic starch micropheres (MSM) for parenteral administration of magnetic iron oxides to enhance contrast in magnetic resonance imaging (M.R.I.) especially by passive targeting of the RES system (spleen & liver). One hour after administration of MSM, all detected radioactivity was seen in the major reticuloendothelial organs. Liver tissues accumulated 84,5 % of the doses whereas 6,5 % of the dose was located within the spleen. The biocompatibility of MSM was confirmed.

Fig. 2 shows the biodistribution of ^{59}Fe labelled MSM and ^{59}Fe labelled degradation products as percent of dose per total organ.

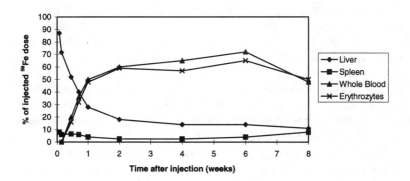

Figure 2 Biodistribution of ^{59}Fe labelled magnetic starch microsphers (MSM) and ^{59}Fe labelled degradation products as percent of dose per total organ. Mean values are plotted.

8 STARCH BASED MICROCAPSULES

Lévy and Andy[10] reported on the production of starch microcapsules using a crosslinking process with terephthaloylchloride applied to hydroxyethyl starch (HES) and sodium starch glycolate. It resulted in stable microcapsules whose size could be adjusted by changing the emulsification parameters. The strength of the wall was shown to depend on polycondensation conditions, mainly pH, terephthaloylchloride concentration and reaction time. HES gave tough membranes, allowing a slow release of encapsulated salicylate. Microcapsules prepared from sodium starch glycolate exhibited hydrophillic properties, as shown by water-induced swelling and gel formation. This special property was optimized by introduction of variability in crosslinking parameters. All crosslinked polysaccharide microcapsules were characterized by a total resistance to digestive media. However, the addition of a protein, such as gelatin, to the starch derivatives in the aqueous phase, provided biodegradable microcapsules.

9 BUCCAL BIOADHESIVE TABLET FORMULATION

In order to increase the buccal residence time of miconazole in cases of oral candidiasis, a bioadhesive buccal tablet with slow release properties has been developed[11]. The main advantages of this delivery system are a reduction in the frequency of administration and in the amount of drug administered, which might improve patient compliance. The bioadhesive tablet formulation contained physically modified maize starch (cold water soluble waxy maize), Carbopol 934, sodium benzoate and silicon dioxide. The powders were blended and directly compressed. The tablets were 2 mm thick and had a diameter of 7 mm. In several studies miconazole nitrate was used as a model drug for local buccal therapy[12-13]. Buccal gels containing miconazole nitrate are currently used for the treatment of e.g. oral candidiasis. They must be applied several times a day. The salivary miconazole nitrate concentrations after administration of the bioadhesive tablet and of oral gels were compared (Fig. 3a & 3b).

Although the amount of drug administered via the bioadhesive tablet was six fold lower than when the gel was used, the salivary miconazole levels were higher and remained above the MIC value of Candida albicans for more than 10 hrs. The mean adhesive time of the tablet was 586 min. The gingiva seemed to be the best site for application of the buccal bioadhesive system[14]. Ongoing experiments indicate that the feasible application of this system for vaginal drug administration.

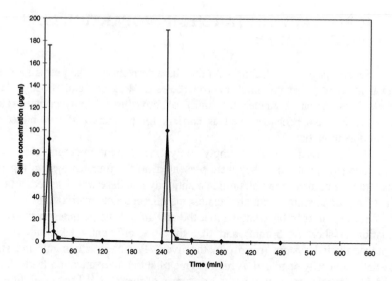

Figure 3a Mean (SD; n = 15) salivary miconazole concentrations after repeated application of 60 mg miconazole nitrate as Daktarin oral gel.

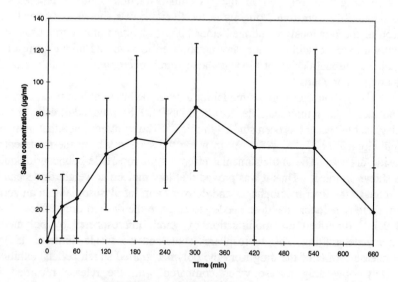

Figure 3b Mean (SD; n = 15) salivary miconazole concentrations after administration of 10 mg miconazole nitrate as a slow-release bioadhesive tablet.

10 POLYACRYL STARCH MICROPARTICLES AS DRUG CARRIERS

Several papers are dealing with the characterization[15], the preparation and the evaluation of polyacryl starch microparticles as drug carriers. Biodegradable microparticles (mean diameter 0.5 mm) of cross-linked polyacrylate starch (maltodextrin) have been designed as carriers for proteins and low molecular weight drugs in vivo.

It was described that empty polyacryl starch microparticles were nonimmunogenic, but when they were presented to the immune system together with entrapped human serum albumin, an antibody response was detected not only against the protein antigen, but also against the microparticle matrix[16].

Studies in mice have shown that the half-life of the particles in the blood circulation is short (< 5 min) and that they are efficiently taken up by the reticuloendothelial system (mainly in the liver). Particles with many and short cross-links are easily degraded. A high degree of starch derivatization leads to less degradable particles remaining in the lysosomes of the RES[17]. The possibilities of using polyacryl microparticles as a carrier for low molecular weight drugs have been investigated. Drugs mainly containing primary amino functions (as e.g.. primaquine) have been coupled covalently to starch microparticles via tri-, tetra-, and pentapeptide spacer arms. The drug-carrier complexes were stable in serum but free drug was released in the lysosomal fraction[18]. Some problems of biocompatibility were found, mainly in the liver[19].It was shown that ester bonds instead of the previously mentioned ether bonds, obtained after derivatisation of starch with acrylic acid chloride, are metabolised in vivo and might form a new basis for a rational design of biodegradable starch microparticles with crosslinks containing ester bonds[20].

Using dinitrophenol as a model for a drug, the humoral immunogenicity of the polyacryl starch microparticles to which DNP has been bound directly or via a biodegradable spacer has been studied in mice[21]. The study revealed that no major immunological obstacles were seen in using the system, e.g. in the treatment of parasitic diseases. The antileishmanial effect of microparticle-bound primaquine was shown in mice[22]. Thus it was proved that low molecular weight drugs can be delivered to the liver macrophages and deliver them intralysosomally in an active form, thereby reducing the dose needed to achieve the desired drug effect. Vyas and Jain[23] described the modification of starch microspheres by polymethyl methacrylate grafting. The microspheres contained isosorbide dinitrate (ISDN). The release profile of the drug from the polymer grafted starch pellets, exhibited relatively slow drug release when compared with the release recorded for polyacrylic acid grafted starch. From in vivo evaluation in rabbits it could be concluded that ISDN could be made available in significantly higher concentrations following buccal administration and eliminate the peaks and valleys of drug plasma profiles associated with a conventional multiple dosing.

11 CHEMO-OCCLUSION WITH DEGRADABLE STARCH MICROSPHERES

The use of particulate embolic agents combined with regional chemotherapy in the treatment of hepatocellular carcinoma and metastatic liver cancer has been widely investigated over the last decade[24-25]. The rationale for the use of such agents is to provide vascular blockade, resulting in a reduced or halted blood flow. This increases the in situ time, tumour exposure and, thus, the effect of any coadministered cytostatic drug. Of all the embolic agents and techniques the degradable starch microspheres (DSM) are the agents that have been evaluated most extensively.

Phase II clinical trials have demonstrated their efficacy when coadministered with chemotherapeutic drugs (chemo-occlusion), as measured by tumour response. The therapeutic benefits associated with chemo-occlusion would suggest that this technique might have a potential application as an adjuvant therapy, e.g. in reducing tumour recurrence after surgical resection in hepatocellular carcinoma, or downstaging a tumour prior to surgical resection, respectively. Furthermore, comprehensive management of patients with liver metastases and potential extrahepatic involvement may well be achieved by a combination of DSM chemo-occlusion and systemic chemotherapy. Large, randomised trials are required to access the clinical benefits associated with chemo-occlusion such as quality of life, time to tumour progression and survival. Other application areas could be the treatment of breast and pancreatic carcinomas. DSM are produced by cross-linking a hydrolysed starch with epichlorohydrin in a common emulsion polymerisation process. The DSM that have been successfully used for chemo-occlusion have a diameter of 45 ± 7 µm. They are readily degraded in the blood by α-amylase, but maintain their integrity for a certain time before they are degraded and disappear from the blood flow. The size of DSM has a direct relationship to the place of occlusion along the vascular tree of the target organ. DSM descend to the arterial capillary level where they are lodged. As the DSM are deformable they have the ability to adapt to their vascular surroundings, so providing a more complete blockade of the capillaries. Approximately 80 % of the blood stream can be temporarily occluded by DSM for a period of 15 - 80 min. A cytostatic drug coadministered with the DSM will consequently be selectively trapped along with the DSM, enhancing concentrations of the drug in the vicinity of the tumour.

12 CONCLUSION

I hope that in this review the prospect of starch as a potential and essential component for drug dosage forms has been demonstrated. The evolution of drug formulation might transpose starch, from its todays position as a simple excipient, into an integral component of effective drug delivery systems with increased bioavailability and biocompatibility.

REFERENCES

1. G. Vandenbossche, R. Lefebvre, G. De Wilde, J.P. Remon - *Performance of a modified starch hydrophillic matrix for the sustained release of theophylline in healthy volunteers.* J. Pharm. Sci. 1992, **87**, 3: 245-248

2. J. Herman, J.P. Remon - *Modified starches as hydrophillic matrices for controlled oral delivery. - In vitro drug release evaluation of thermally modified starches.* Int. J. Pharm. 1989, **56**: 66-70.

3. K. Ridgway - *Hard Capsules - Development and Technology.* The Pharmaceutical Press 1987. B.E. Jones - Chapter 4 - *Gelatin Additives, Substitutes and Extenders.* Page 58.

4. H. Te Wierik - *Preparation, characterisation and pharmaceutical application of linear dextrins.* Thesis Rijksuniversiteit Groningen (Ned.), 1993.

5. L. Illum, N. Farraj, M. Critchley, S.S. Davis - *Nasal administration of gentamicin using a novel microsphere delivery system.* Int. J. Pharm. 1988, **46**: 261-265.

6. E. Björk, P. Edman - *Degradable starch micorspheres as a nasal delivery system for insulin.* Int. J. Pharm. 1988, **47**: 233-238.

7. N.F. Farraj, L. Illum, S.S. Davis, B.R. Johansen - Diabetologia 1989, **32**: 486A.

8. A. Fahlvik, E. Holtz, U. Schröden, J. Klaveness - *Magnetic starch microspheres biodistribution and biotransformation.* Invest. Radiol. 1990, **25**: 793-797.

9. P.G. Krishnan, I.T. Julson, D.J. Robinson, Y.V. Pathak - *Polyethylene-starch extrudates as erodible carriers for bioactive materials.*Advances in Controlled Delivery of Drugs, 1994: 59-79. Ed. by Melvyn A. Kohudic, (Technomic Publishing Co. - Lancaster - Basel.

10. M.C. Lévy and M.C. Andy - *Microcapsules prepared through interfacial crosslinking of starch derivatives.* Int. J. Pharm. 1990, **62**: 27-35.

11. S. Bouckaert and J.P. Remon - *In vitro bioadhesion of a buccal, miconazole slow release tablet.* J. Pharm. Pharmacol 1993, **45**: 504-507.

12. S. Bouckaert, M. Schautteet, R.A. Lefebvre, J.P. Remon, R. Van Clooster *Comparison of salivary miconazole concentration after administration of a bioadhesive slow-release buccal tablet and an oral gel.* Eur. J. Clin. Pharmacol 1992, **43**: 137-140.

13. S. Bouckaert, R.A. Lefebvre, J.P. Remon - *In vitro - in vivo correlation of the bioadhesive properties of a buccal bioadhesive miconazole slow-release tablet.* Pharm. Res. 1993, **10**: 853-856.

14. S. Bouckaert, R.A. Lefebvre, F. Colardyn, J.P. Remon - *Influence of the application site on bioadhesion and slow-release characteristics of a bioadhesive buccal slow-release tablet of miconazole.* Eur. J. Clin. Pharmacol. 1993, **44**: 331-335.

15. P. Artursson, P. Edman, T. Laakso, I. Sjöholm - *Characterization of polyacryl starch microparticles as carriers for proteins and drugs.* J. Pharm. Sci. **73**: 1507-1513.

16. P. Artursson, I.L. Martensson, I. Sjöholm - *Biodegradable microspheres : some immunological properties of polyacryl starch microparticles.* J. Pharm. Sci. 1986, **75**, 697-701.

17. T. Laakso, P. Artsursson, I. Sjöholm - *Biodegradable microspheres : Factors affecting the distribution and degradation of polyacryl starch microparticles.* J. Pharm. Sci. 1986, **75**, 962-967.

18. T. Laakso, P. Stjärnkvist, I. Sjöholm - *Lysosomal release of covalently bound antiparasitic drugs from starch microparticles.* J. Pharm. Sci. 1987, **76**: 134-144.

19. T. Laakso, P. Edman, U. Brunk - *Biodegradable microspheres : Alterations in mouse liver morphology after I.V. administration of polyacryl starch microparticles with different biodegradability.* J. Pharm. Sci. 1988, **77**: 138-144.

20. T. Laakso & I. Shöholm - *Biodegradable microspheres : some properties of polyacryl starch microparticles prepared from acrylic acid-esterified starch.* J. Pharm. Sci. 1987, **76**: 935-939.

21. P. Stjärnkvist, L. Deyling, I. Sjöholm - *Biodegradable microspheres: Immune response to the DNP hapten conjugated to polyacryl starch microparticles.* J. Pharm. Sci. 1991, **80**: 436-440.

22. P. Stärnkvist - *Biodegradable microspheres : Effect of microparticle bound primaquine on L. donovani in mice.* Int. J. Pharm. 1993, **96**: 23-32.

23. S.P. Vyas, C.P. Jain - *Bioadhesive polymer-grafted starch microspheres bearing isosorbide dinitrate for buccal administration.* J. Microencapsulation 1992, **9**: 457-464.

24. P. Taguchi - *Chemo-occlusion for the treatment of liver cancer.* Clin. Pharmacokin. 1994, **26**: 275-291.

25. A.B.S. Ball - *Regional chemotherapy for colorectal hepatic metastases using degradable starch microspheres.* Acta Oncologica 1991, **30**: 309-313.

26. V. Lenaerts, Y. Dumoulin, M.A. Meteesen - *Controlled release of theophylline from cross-linked amylose tablets.* J. Control. Rel. 1991, **15**, 39-46

Trehalose and Novel Hydrophobic Sugar Glasses in Drug Stabilization and Delivery

E. M. Gribbon, R. H. M. Hatley, T. Gard, J. Blair, J. Kampinga and B. J. Roser

QUADRANT HOLDINGS CAMBRIDGE LIMITED, MARIS LANE, TRUMPINGTON, CAMBRIDGE CB2 2SY, UK

INTRODUCTION

Ideal drug delivery is the effective and convenient administration of a prophylactic or therapeutic agent in a precise dose to the specific target tissue at the correct time. This is rarely achieved using conventional oral dosage forms which require repeat dosing at specific time intervals, resulting in fluctuating blood levels. Development of any new or modified drug entity to overcome these problems demands heavy investment in both time and money. More stringent control over drug testing by the regulatory authorities has also resulted in a decrease in the number of new products coming onto the market annually. This narrows the time-window to exploit proprietary compounds.

These factors have contributed to the deployment of substantial resources into optimising drug therapy by improved drug delivery systems. The pharmacology, chemistry, therapeutic indications and side effects, being already well documented for existing drugs, encourage the concentration of effort into development of novel forms of drug delivery rather than novel drugs themselves. Appropriate delivery systems can extend the duration of action for those drugs with short half-lives, can reduce side effects and result in better patient compliance.

Progress in the development of novel controlled release materials and methods for effectively incorporating stable pharmaceuticals in them have been the two most recent significant areas to have enlivened the field of drug delivery. The Q-T4 drying technology and Quadrant's portfolio of novel slowly soluble glass bio-materials have a significant contribution to make in the development of effective, novel drug delivery systems.

Trehalose stabilisation: Following the lead of Nature

Quadrant's Q-T4 drying technologies were primarily focused on harnessing the unique features of the natural phenomenon known as *cryptobiosis,* literal meaning: "hidden life". Cryptobiotic species are widely distributed throughout both the plant and animal kingdoms, ranging from the "resurrection plant" *Selaginella lepidophylla* to brine shrimps *Artemia salina* [1]. They all share the attribute of undergoing complete physiological desiccation under harsh conditions and then remaining in this dried, and apparently dead, state indefinitely. On subsequent exposure to water however, they fully rehydrate and come back to life. When in the dried state, they have the capacity to withstand prolonged exposure to environmental extremes including temperatures as great as 70 °C [2].

It was established early that a common feature amongst cryptobionts was the presence, in high concentration, of the simple yet unusual disaccharide trehalose [3]. This is comprised of two molecules of glucose linked via their reducing carbon atoms. Hence, trehalose is non-reducing. It is also non-toxic [4]. A simple illustration of this lies in the fact that many common cryptobionts contain up to 20% trehalose and are everyday foodstuffs. A familiar example is yeast *Saccharomyces cerevisiae* contained in bread and fermented products such as beer and wine [5]. Once ingested, trehalose is rapidly broken down to its constituent glucose molecules by trehalase, a specific enzyme found in the gut and circulation of virtually all omnivorous species including man [6].

The initial working hypothesis, that trehalose could stabilise all labile entities *in vitro* and on years of storage at elevated temperatures has been repeatedly confirmed both in the laboratory and in the marketplace [5,7]. The outstanding feature of these technologies is that they mimic cryptobiosis, in that they occur at ambient temperatures and are consequently very fast compared to the antiquated freeze-drying method widely adopted within the pharmaceutical industry.

Limitations of Freeze-Drying

The freeze-drying process is currently standard practice and is approved by all regulatory authorities. Nevertheless it has a number of intractable disadvantages. The first of these is the freezing stage itself, which results in freeze-concentration. As ice crystals form within the freezing solution, the solute concentrations within the residual phase of unfrozen water increase [8,9]. These increases may have harmful effects such as greatly increased reaction kinetics for those solutes that are most reactive in solution. An example concerns a commercially available kit for the measurement of haemoglobin in blood [10]. The assay solution containing potassium ferricyanide and potassium cyanide salts, is quite stable under refrigeration. However, on freezing, the ferricyanide ion is reduced to the ferrocyanide form and the assay does not work. Likewise, the colloidal hydrogels of the hydroxide and phosphate salts of aluminium, the sole FDA approved adjuvants available for use within the vaccine industry, are completely inactivated if frozen [11]. The other great danger associated with freeze-concentration is precipitation or crystallisation of components such as buffer salts which reach saturation during freeze-concentration. In this instance there are pH deviations which are often sufficient to denature any fastidiously pH dependent product [12].

Because the whole primary drying process occurs by the sublimation of water vapour from ice crystals at low temperatures, freeze-drying is inherently slow. This is compounded by the fact that the secondary drying stage involves the evaporation of water vapour from the residual amorphous glass phase formed during freeze concentration. Water molecules have very limited mobility in glasses so that this stage of the process is also slow especially at low temperatures[13].

The Glassy State and Glass Transition Temperature

The stability of any product dried in trehalose is based on the glassy amorphous state. Solids which exist in the crystalline state have a defined, ordered structure. However, those which exist in the amorphous or glassy state have a generally disordered structure. Trehalose, like any other carbohydrate generally exists in the crystalline state. However, if the crystalline material is melted and subsequently cooled at a rapid rate which does not permit nucleation to occur, solution viscosity rises until, over a narrow specific temperature range, the viscosity rapidly increases and the carbohydrate becomes a glass. This is referred to as the glass transition temperature (Tg) for that particular material. The high viscosity confers stability to the glass by retarding the reaction kinetics of chemical processes.

Some carbohydrates may also form glasses from solution. The removal of water molecules may be by either freezing the solution and subliming the ice as in freeze-drying or by evaporation. The Tg of the final dry product is a function of the residual moisture content. The greater the residual moisture content, the lower the Tg will be. Trehalose, when added to a product as a stabiliser, not only aids in the glass-forming process but helps to protect the active material from the detrimental effects of concentration.

The Tg was previously thought to represent a critical temperature above which the rate of deterioration of an active product is rapid and should therefore never be exceeded when the product is stored. It was assumed that storage below Tg was safe. However, it has recently been shown that degradative reactions can occur below Tg. One example of this is the stability of catalase which does not relate to the glass transition temperature of the stabilising excipient [14]. Another illustration is the poor storage stability of dried human growth hormone incorporated in a dextran glass which itself has a high glass transition temperature.

The mechanisms for these degradative reactions have not been fully elucidated. It has been postulated by others that they may originate from side chain flexibility of the protein and/or diffusion of small molecules such as water or oxygen through the glassy matrix, resulting in deleterious hydrolytic or oxidative reactions [15]. However, there is a close correlation between instability of dried biomolecules and the inherent chemical reactivity of the glass-forming excipient itself. This strongly implicates chemical reactions such as the Maillard reaction between the active and the excipient as the major cause of product instability [16,17]. There is also clear evidence that very labile products can exhibit remarkable stability in trehalose even when heated above the Tg [7].

Q-T4 Drying and Stabilisation in Trehalose

Freeze-drying, involving extremely low temperatures, is a very slow process, occurring over periods of days and is therefore expensive. However, any product formulated with a glass-forming excipient such as trehalose, may be dried directly from solution, using controlled evaporative drying, thereby eliminating the need for freezing [18,19]. A wide variety of drying techniques can be used including spray, tray, drum or vacuum drying. Minimum changes to validated GMP processes can be achieved by using existing installed freeze-dryers with modified run parameters to disable the freezing step and to give controlled ramping of temperature and vacuum. The degree of denaturation associated with the drying of any product, is a function of the time taken, the temperature of the product prior to achieving the glassy state and the chemical reactivity of the components of the formulation, especially the stabilising excipient. Controlled evaporative drying can hold the product at cool temperatures, without freezing, prior to the establishment of the glassy state. This not only has the advantage of permitting the drying of freeze-sensitive materials but is fast; drying to residual moisture contents of < 1% being complete in a few hours.

Based on their glass forming ability, many sugars have been advocated for ambient temperature drying. However, in addition to forming a glass, the carbohydrate must also be chemically inert and not react with the product. Reducing sugars, by their nature, are unsuitable, since deleterious reactions, as typified by non-enzymatic browning (the Maillard reaction[18]) frequently occur. Sucrose, itself non-reducing, is easily hydrolysed in any drying solution, to yield reducing sugars (Fig 1). Chemical instability, especially in the presence of reactive products, renders most other non-reducing carbohydrates equally unsuitable for producing robust dried pharmaceuticals.

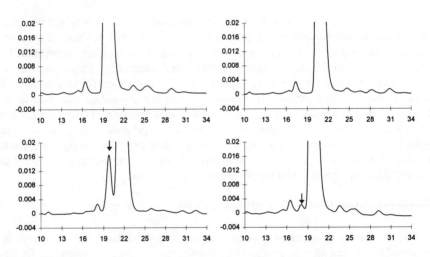

Figure 1   ***Reverse Phase HPLC of glucagon after drying and storage***

TOP LEFT   Fresh glucagon

TOP RIGHT   Trehalose dried and stored at 60 °C for 2 weeks. There is no change fom the fresh material

BOTTOM LEFT   Glucose dried. Immediately on drying there is a new peak of chemically altered glucagon which runs just ahead of the main peak (↓). This is glycated glucagon, a Maillard reaction product. Within a further 4 days at 60 °C all the glucagon was glycated and inactive (not shown).

BOTTOM RIGHT   Sucrose dried and stored at 60 °C for 2 weeks. There is a small new peak of glycated glucagon running in the same place as in the glucose sample (↓). This degradation process is also progressive and leads to loss of most of the glucagon within 2 months (not shown).

Since the glassy state is an essential prerequisite to stability, it is also crucial that the selected carbohydrate should not crystallize during drying. The majority of sugars exhibit this deleterious attribute; *e.g.* mannitol and lactose usually fail to confer room temperature stability because they crystallise so readily. Trehalose is an excellent glass former and the resulting glasses interact with water vapour in a unique way so that only the surface of the glass absorbs water and recrystallises while the body of the glass resists further water uptake. It is also chemically inert making trehalose in combination with Q-T4 drying techniques the carbohydrate of choice for product stabilisation.

Q-T4 Stabilisation and Novel Drug Delivery Systems

A wide range of pharmaceuticals have been successfully dried in trehalose using Quadrant's Q-T4 drying technologies. Examples range from recombinant peptides such as human growth hormone, glucagon and Factor VIII, to vaccines including multivalent vaccines. Furthermore, many of these contain aluminium hydroxide or phosphate as the adjuvant, which has also been successfully stabilised in the dry state. Since this adjuvant cannot be frozen or dried by other techniques, existing formulations require transportation at 4°C resulting in exorbitant cold-chain costs [21].

The trehalose glass is in the form of a fine-textured solid foam in which the active is immobilized in solid solution in the thin glassy walls of the foam bubbles., Some freeze-dried materials may take 5 minutes or more to fully dissolve when reconstituted. In contrast, all Q-T4 dried materials dissolve instantly and completely. This foam technology may be taken one stage further to produce a basic first generation delivery format. This eliminates the need for a separate sterile vial containing the dried drug. The material in the pre-filled syringe is rehydrated immediately prior to administration. Presentation of a drug in this format, using the Q-T4 approach, does not add to production costs, as any type of currently available syringe may be used. Although other pre-filled syringes are available that contain a limited range of freeze-dried products, they are much more expensive since specialised syringes are required for the freeze-drying process.

Utilisation in Conventional Slow Release Systems

Many current slow release systems require that the actives for incorporation be thermostable and resistant to certain organic solvents [22]. This automatically limits the number of active materials that may be considered for development. Quadrant, in addition to successfully stabilising a wide range of products for high temperature storage, have shown that trehalose-dried actives are completely resistant to denaturation by all those organic solvents in which trehalose itself is insoluble. For example horse radish peroxidase and alkaline phosphatase dried in trehalose are resistant to exposure to a wide range of organic solvents including chloroform, acetone, xylene, dichloromethane and tetrachloroethylene, for periods of at least 72 hours. Using recombinant hepatitis B surface antigen, it was also recently shown that resistance to solvents such as dichloromethane and ethylacetate was conferred by Q-T4 drying, enabling solvent-mediated incorporation into polylactide / glycolide microspheres, as well as other novel systems for the development of a pulsatile release vaccine.

Controlled Release Technology

The stable incorporation of pharmaceuticals into a glassy matrix obviously creates a specific type of delivery system in itself. Novel, biocompatible, non-toxic, synthetic glasses which are surface eroded by phase transitions at slow and controlled rates have recently been developed. An active, either in solid solution or suspension in disks of these matrices are automatically released in a zero order fashion (Fig 2) as the erosion front moves through the glass. Because these glasses are inherently hydrophobic, they do not swell in body fluids by imbibition of water. The product therefore remains dry and stable until reached and released by the erosion front. An extensive list of these glasses has been synthesised and they are being developed to produce controlled release systems with characteristics suitable for a wide range of applications. Predetermined release rates can be precisely controlled by chemical design of the biomaterial and the physical design of the delivery system.

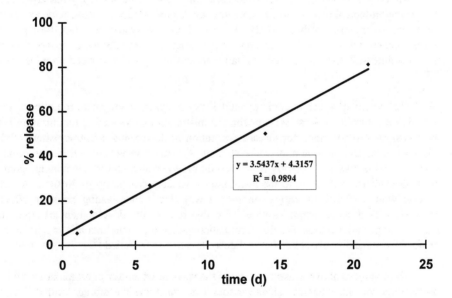

Figure 2   *In vitro release of synthetic corticosteroid from hydrophobic glass.*

The experimental synthetic corticosteroid XPDO was dissolved in a melt of the modified sugar hydrophobic glass TOAC at 10% w/w and of 2.5mm glass beads were cast. These were assayed for product release into phosphate buffered saline pH 7.4 at 37°C with frequent buffer changes. Note that the release is linear with time ("zero order". $R^2 = 0.99$) at the rate of about 3.5% of the total dose per day over 3 weeks.

Most of the standard polymeric delivery systems produce first order drug release; the rate of release being dependent on the concentration of drug remaining in the delivery system. Ideally, drug release should be zero order as in these novel hydrophobic glasses so that in clinical practice, the amount of drug reaching the systemic circulation is constant and independent of the amount of drug remaining in the dosage form. A steady-state condition ensures a constant drug level in order to achieve the desired therapeutic response[23].

Novel Drug Delivery Systems and the Future

Most currently available long-term controlled release matrices are made from non-degradable implantable material. One example is Norplant, a 5 year sub dermal implant for controlled release of levonorgestrel for contraception [24]. These silastic rods have the disadvantage of requiring surgery for removal as well as insertion. Furthermore, only diffusional rather than erosional release is available[22] and some questions remain concerning their toxicity especially to the immune system.

Although systems such as liposomes are being studied, the difficulty has always been preserving biological activity while achieving the desired release kinetics. One example is the work, described by Kibat *et al* 1990 [25], of a liposome system developed for pulsatile release and encapsulated within calcium alginate microspheres. Although encapsulation of the liposomes stabilised the protein initially, activity was greatly reduced after 30 days at 37°C.

The logical approach to commercial delivery systems development might be to focus initially on controlled release systems for administration of vaccines. Immunisation is the most cost effective weapon for disease prevention in developing countries where 80% of the world's population live and where 86% of all births and 96% of all deaths occur [26]. However, immunisation currently requires multiple injections and in developing countries drop-out rates from the first to the last doses of vaccine can be up to 70% [27]. An ideal vaccine would deliver the antigen in such a way that a long-lasting boosting effect is achieved with a single administration. For this reason, the development of controlled-release and pulsatile-release vaccine formulations is an important area and in spite of a lot of work the standard polylactide / glycolide system has so far failed [28].

The examples above illustrate a few of the ways in which novel biomaterials with the useful physical and chemical characteristics can transform the storage and delivery of drugs. Quadrant, with its Q-T4 and novel glass technologies can provide previously attainable opportunities in novel drug delivery.

REFERENCES

1.   Clegg, J.S. "Dry biological systems" Academic Press, London 1978 Crowe, J.H. and Clegg, J.S. ed pp 117

2.   Leopold, A.C. 1986 "Membranes, metabolism and dry organisms".Cornell University Press, Ithaca and London

3.   Crowe, J.H. and Crowe, L.M. 1986 "Membranes, metabolism and dry organisms" Cornell University Press, Ithaca and LondonLeopold, A.C. ed pp 188

4. U.S. F.D.A.:Drug Master File Type 5 "Trehalose Stabiliser for Biological Molecules" 1994

5. Roser, B. *Trends in Food Science and Technology* 1991 **2** 166

6. Takesue, Y., Ikezawa, H., Nishi, Y., Taguchi, R and Yokota, K. 1986 *Cell Structure and Function* **11** , 491

7. Colaco, C., Sen, S., Thangavelu, M., Pinder, S and Roser, B.*Biotechnology* 1992 **10**, 1007

8. Hatley, R.H.M. and Franks, F. *J. Thermal Analysis* 1991 **37**, 1905

9. Pikal, M.J. 1990 *Biopharm* **3**, 18

10 Hatley, R.H.M., Franks, F., Day, H. and Byth, B. 1986 *Biophysical Chemistry* **24**, 41

11 Nail, S.L., White, J.L. and Hem, S.L. 1976 *J. Pharm. Sci* **65**, 1188

12. Murase, N. and Franks, F. 1989 *Biophys Chem* **34**, 293

13. Hatley, R.H.M., Franks, F. and Day, H. 1986 *Biophys Chem* **24**, 187

14. Kerr, W.L., Lim, M.H., Reid, D.S. and Chen, H. 1993 *J.Sci. Food Agric.* **61**, 51-56

15. Roy, M.L., Pikal, M.J., Rickard, E.C. and Maloney, A.M. 1992 *Devel in Biol Standard.* **74**, 323

16. Carpenter, J.F., Prestrelski, S.J., Anchordoguy, T.J. and Arkawa, T. 1994.: "Formulations and delivery of proteins and peptides" American Chemical Society Symposia Series No. 567 Cleland J. & Langer, R eds... pp 134

17 Colaco C.A.L.S., Smith C.S., Sen S., Roser D.H., Newman Y., Ring S. and Roser B.J.."Formulation and Delivery of Proteins and Peptides" American Chemical Society. J.L. Cleland and R. Langer eds. 1993 pp 221

18. Franks, F. and Hatley, R.H.M. " Stability and Stabilisation of Enzymes" Elsevier Holland 1992 Van den Tweel, W.J.J., Harder, A. and Buitelaar, R.M. eds pp 45-54,.

19. Franks, F., Hatley, R.H.M. and Mathias, S.F. 1991 *Biopharm*, **14**, 38

20. "The Maillard reaction in food processing, human nutrition and physiology" Birkhauser, Basel. 1990. Finot P.A., Aeschbacher H.U., Hurrell R.F. & Liardon R. eds..

21. Beardsley, T. 1995 *Scientific American* **272** 1, 68

22. Sanders, L.M. 1991 "Peptide and protein drug delivery". Marcel Dekker Inc., New York Lee, V.H.L. ed, pp 785

23. Rubinstein, M. 1993. "The First European Pharm. Tech. Conference". Advanstar Communications, Chester, UK Rosser, M. ed, pp 346

24. Roy, S., Mishell, D.R., Jr, Robertson, D.N. *et al.* 1984 *Amer J Obstes and Gynae*, **148** 7, 1006

25. Kibat, P.G., Igari, Y., Wheatley, M.A., Eisen, H.N. and Langer, R. 1990.*FASEB* **4**, 2533

26. Bloom, B.R. 1989 *Nature*, **324**, 115

27. Aguado, M.T. and Lambert, P.H. 1992 *Immunobiology* **184**, 113

28. Heller, J. 1993. *Advanced Drug Delivery Reviews* **10**, 163

Aqueous Shellac Solutions for Controlled Release Coatings

Manfred Penning

KAISER FRIEDRICH PROMENADE 28A, D-61350 BAD HOMBURG, GERMANY

Introduction

Shellac has been used for pharmaceutical and controlled release coatings for a long time. It was mainly applied from alcoholic solutions which are also known as Pharmaceutical Glazes. Application from various aqueous systems is also possible.

Alkaline shellac solutions have excellent film forming properties, are easy to prepare and apply and show an improved performance in their mechanical and release characteristics even after extended storage.

Shellac

Shellac is the purified product of the natural polymer Lac, the resinous secretion of the tiny insect *Kerria lacca*. India and Thailand are the primary sources of shellac. The properties of shellac depend not only on insect strain and host tree but also to a large extent on the process used for refining.

There are three quite different processes used for the refining of shellac, a melting filtration process, a bleaching process with sodium hypochlorite and refining by solvent extraction.

In this last process, carefully selected and tested crude lac (seedlac) is dissolved in alcohol, preferably ethanol. The batch size of the 10 - 20% solution can be 5000 l or more. The solution is filtered to remove impurities and wax and decolorized by a treatment with activated carbon The solvent is evaporated and the hot liquid resin drawn to a film. After cooling the shellac film breaks into thin flakes. The flakes can be crushed or milled to a fine powder.

Of the three processes used, the refining by solvent extraction is a gentle process for the production of high quality shellac with a narrow specification and a uniform and stable quality required for pharmaceutical purposes. The chemical structure of shellac is not changed by this process.

Shellac refined by solvent extraction can meet the requirements of the *United States Pharmacopoeia USP 23 NF 18 and Deutsches Arzneibuch DAB 10*. It is also approved for food applications as a coating for confectioneries and fresh fruits by the EC and has a GRAS status by the FDA.

Due to a low melting range, shellac can also be applied from hot melts. Microencapsulation of liquids by the coacervation process is also possible.

Alcoholic Shellac Solutions

Shellac has been used since 1930 as an enteric coating material from alcoholic solutions. Problems with a certain polymerization and hardening after storage are well known. This is due to a polycondensation and esterification of the carboxyl groups with alcohol groups.

Several methods are known and used to stabilize alcoholic shellac coatings. Blocking of the carboxyl groups with polymers that contain free amino groups, e.g. poly-vinylpyrrolidone, polymerization to an equilibrium by tempering the coated tablets, mixtures of shellac with HPMCP, etc.

By using an ethanolic shellac solution without plasticizer or additives (Formulation 3) a change of the release properties after a storage of one year is clearly noticeable (Figures 5 and 6).

Aqueous Shellac Systems

Several methods are known to prepare aqueous shellac systems for pharmaceutical coatings. Aqueous-alcoholic solutions, shellac emulsions, pseudolatex dispersions with or without residual solvents and shellac suspensions for filmforming by thermogelation are described in the literature. Most of these systems still contain organic solvents to a certain extent or require inlet and product temperatures well above the MFT (Minimum Filmforming Temperature) of the polymer.

Aqueous Shellac Solutions

Shellac, like other polymers with carboxyl groups, is not soluble in water. However, it is possible to prepare aqueous shellac solutions of alkali salts. The selection of the base and the method for dissolving will influence the properties of the film. A volatile alkali is preferable. Therefore ammonium carbonate was chosen as the base.

The aqueous solutions were prepared by dissolving shellac in demineralized water with ammonium carbonate under stirring and heating. At higher temperatures (<50 °C) the formation of CO_2 and NH_3 occurred; both compounds are volatile. Therefore excessive ammonium carbonate which was not used for the ammonium salt formation of shellac evaporated from the solution. The pH of the clear shellac solution was 7.3 - 7.35. After cooling to room temperature a plasticizer (5 - 10% w/w, on polymer) can be added.

Shellac 2011, a micronized shellac refined by solvent extraction, was used in the following formulations:

Figures

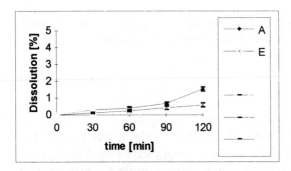

Fig 1: Dissolution of various coated paracetamol pellets, pH 1 (0,1n HCl); A= aqueous shellac solution without plasticizer, E= ethanolic shellac solution

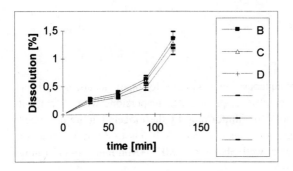

Fig 2: Dissolution of various coated paracetamol pellets , pH 1 (0,1n HCl); B= aqueous shellac solution with Triacetin, C= aqueous shellac solution with propylenglycol, D= aqueos shellac solution with Citroflex 2

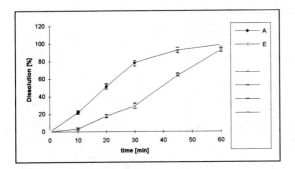

Fig. 3: Dissolution of various coated
paracetamol pellets, pH 6,8 (phosphate
buffer)

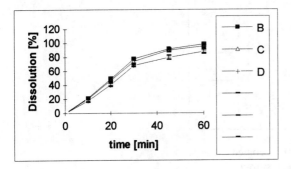

Fig. 4: Dissolution of various coated
paracetamol pellets, pH 6,8 (phosphate
buffer);

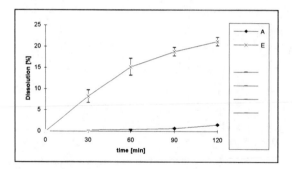

Fig. 5: Dissolution of various coated paracetamol pellets, after one year storage (25°C, 60% humidity) pH 1

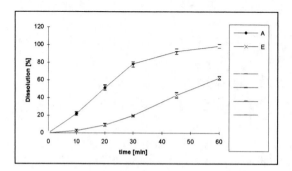

Fig. 6: Dissolution of various coated paracetamol pellets, after one year storage, pH 6,8 (phosphate buffer)

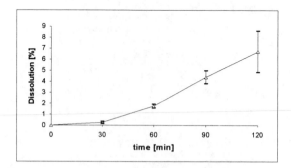

Fig. 7: Dissolution of paracetamol pellets ccoated with an aqueous shellac solution , Triacetin 5%, Talcum 25% (phosphate buffer) pH 4,5

Formulation I

Shellac 2011	190.0 g
Ammonium carbonate (dried)	11.7 g
Demineralized water	1670.0 g
Plasticizer	9.5 g

Formulation 2

Shellac 2011	160.0 g
Ammonium carbonate (dried)	9.9 g
Demineralized water	1540.0 g
Triacetin	8.0 g
Talcum	(16.0) 40.0 g

For comparison tests an ethanolic solution of shellac was prepared as follows:

Formulation 3

Shellac 2011	190.0 g
Ethanol	1710.0 g

Coating process

Paracetamol pellets were coated using a Hüttlin ball coater, a special fluid bed coater. The coating conditions for the aqueous solution of shellac were: inlet temperature 50 °C, product temperature 38 - 40 °C, spray rate 13.0 g/min, 1600 g paracetamol pellets.

Coating conditions for the ethanolic solution of shellac: inlet temperature 25 °C, spray rate 10.0 g/min, 1600 g paracetamol pellets. After the coating the pellets were dried at 40 °C for 30 min and cooled down to 25 °C.

Dissolution Tests and Results

Dissolution tests were carried out using the paddle stirring element described in the U.S. Pharmacopeia USP 22 (900 ml medium, temperature: 37+/- 1 °C, 100 rpm).

In the coating tests, the same amount of polymer was coated on the same amount of paracetamol pellets (Figures 1 to 6).

The dissolution tests at different pH values are shown in Figures 1 to 6. In simulated gastric juice the dissolution of the drug from the pellets coated with aqueous shellac was similar to that with the ethanolic coating (Figure 1).

In the acid solution, the ammonium salt of shellac is obviously changed to the protecting shellac acid. The shellac ammonium salt reacted quickly with the acidic medium to shellac acid and ammonium ions.

In a phosphate buffer (pH 6.8), drug release from the pellets coated with the aqueous shellac solution is significantly faster compared to those coated with the ethanolic shellac solution (Figure 3). Obviously the ammonium salt of shellac dissolved faster than the free shellac acids from the ethanolic shellac solution.

Triacetin and propylene glycol seemed to be suitable as plasticizers (5 - 10% to the amount of polymer w/w). In order to improve the drug release in phosphate buffer pH 6.8 the amount of shellac coated onto the pellets was reduced (Formulation 2).

In Figure 5 the improved release of paracetamol is shown after reducing the film thickness from about 3.5 mg/cm^2 to 3 mg/cm^2. The dissolution in 0.1N HCl did not change significantly.

In an acidic medium the coated paracetamol pellets had a tendency to stick to each other, and therefore talcum was used (Formulation 2).

The dissolution of paracetamol in a phosphate buffer at pH 4.5 after one hour was less than 3% and after 2 hours less than 8%: see Figure 7.

After one year of storage (at 25 °C and 60% RH) the dissolution rate from the ethanolic shellac coated pellets in simulated gastric juice is higher than immediately after the coating process (Figure 5).

In phosphate buffer (pH 6.8) the dissolution rate of these pellets decreased after storage (Figure 6).

The drug release of the pellets coated with the aqueous shellac solution did not change. It is believed that a polymerization or a hardening of the shellac film was prevented by changing the shellac acid into the ammonium shellac salt. Storage for one year had no influence on the drug release.

Mechanical Strength

It is well known that the mechanical strength of films cast from aqueous shellac solutions is higher than those from alcoholic solutions. This is probably due to the salt structure of the aqueous films compared to a more amorphous character of the films from the ethanolic solution.

Preparation and Testing

The aqueous and alcoholic shellac solutions were prepared as descibed above.

The formulations were prepared with and without plasticizer (10% w/w triacetin on resin). The solutions were cast in petri dishes and dried at 40 °C to a residual solvent content of approximately 1.6%. The films were isolated and cut into pieces of 30 x 6 mm. The elasticity of the films was measured by a Texture Analyser TA-XT2.

Type of Film	Film thickness (μm)	Res. Solvent (%)	Tensile Strength (N)	Elongation (mm)
Film ASL, 5%	200 - 220	1.6	21.7	0.26
Film ASL, 10%	200 - 220	1.4	41.5	0.83
Film ESL	150 - 170	1.7	9.56	0.21
Film ESL, 10%	150 - 170	1.8	4.83	0.74

Table 1: Film ASL, 5% = Aqueous shellac solution with 5% plasticizer
Film ESL, 10% = Ethanolic shellac solution with 10% plasticizer

Even so the absolute results are not directly comparable due to the difference in film thickness. The films from the aqueous shellac solution showed a significant increase in tensile strength with increased amount of plasticizer, whereas the films from the ethanolic shellac solution showed a reduction in tensile strength with increased plasticizer content.

Summary

The selection of the grade of shellac is important for a stable and uniform quality and the performance of the polymer film. The molecular structure of shellac refined by solvent extraction is not changed during the refining process.

The application of aqueous shellac solutions for enteric and other pharmaceutical coatings will not only avoid the problems with organic solvent systems but also improve the performance of the polymer film by stable dissolution characteristics after extended storage time and result in improved mechanical properties compared to films from ethanolic shellac solutions.

References

Bauer, K.H. and Osterwald, H.P., Studien über wäßrige Applikationsformen einiger synthetischer Polymere für dünndarmlösliche Filmüberzüge. Pharm. Ind. 12 (1979) 1203-1209.

Bose,P.K.,Sankaranarayanan,Y., Sen Gupta,S.C.,Chemistry of Lac,Indian Lac Research Institute ,Ranchi ,1963

Bueb, W., Untersuchung und Charakterisierung eines neuen Überzugsverfahrens in Verbindung mit der Verarbeitung neuartiger,thermogelierbarer Überzugsmaterialien PhD Thesis ,Albert-Ludwigs-Universität,Freiburg,1993

Chambliss, W.G., The forgotten dosage form : Enteric-coated tablets. Pharm.Technol. 9 (1983) 124-140.

Chang, R.K., Hsiao, C.H. and Robinson, J.R., A review of aqueous coating techniques and preliminary data on release from a theophylline product. Pharm. Technol., 3 (1987) 56-68.

Chang, R.K., Iturioz,G. and Luo, C.-W., Preparation and evaluation of shellac pseudolatex as an aqueous enteric coating system for pellets. Int. J. Pharm., 60 (1990) 171-173.
El Banna, H.M., Efimova, L.S. ,The Construction and Use of Factorial Design in Fluidized Bed Microencapsulation , Pharm.Ind. 44, Nr 6, (1982) 641 - 645

Itoh, S. ,Koyama, H., Hirai, S., Kashihara, T., Enteric Film and preparation thereof. J Pat. 243542/88 (1988)

Johnston, G.W., Malani, R.I., Scott, M.W. , Process for stabilizing shellac coating. US Pat. 3,274,061 , (1966)

Luce, G.T., Disintegration of tablets enteric coated with CAP. Manuf. Chem. Aerosol News, 49 , 1978, (7), 50,52 ,67

Penning, M., Schellack - ein "nachwachsender Rohstoff" mit interessanten Eigen-schaften und Anwendungen .Seifen-Öle-Fette-Wachse 6(1990) 221-224

Signorino, C.A., Jamison, T.E., Coated Tablet , US Pat. 3,576,663 (1971)

Specht, F., Pseudo-Latex Verfahren zur Herstellung von Mikropartikeln und Filmen PhD Thesis,Christian-Albrechts-Universität, Kiel ,1995

Information Requirements for Drug Delivery Systems

Kassy Hicks

DERWENT SCIENTIFIC AND PATENT INFORMATION, DERWENT HOUSE, 14 GREAT QUEEN STREET, LONDON WC2B 5DF, UK

Abstract

The development of appropriate drug delivery systems is an increasingly important part of the drug development process. The cost of developing new drugs is increasing and the decision to enter new markets and create new products and formulations is critical in today's competitive market place.

It is also imperative to be able to monitor the competition and stay ahead of industry trends. Top quality information is essential for decision-making and for innovative R & D and helps to identify opportunities and reduce unnecessary risks.

Major pharmaceutical and scientific companies have long recognized the value of patent and scientific information as a key business tool. In order to try to meet this demand for information, Derwent is currently investigating a new service aimed at pharmacists and pharmaceutical scientists working in drug development formulations and pharmacy.

This paper will discuss the value of patent and scientific information in this area, and focus on the progress of the new service following extensive market research in early 1996.

Subject Index